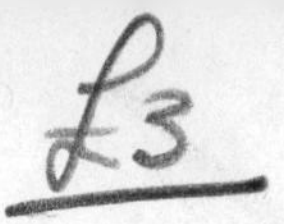

GW00691876

IMAGES

The Wykeham Science Series

General Editors:

The Author:

CHARLES A. TAYLOR worked in the Admiralty during the War, and then joined the University of Manchester Institute of Science and Technology. Since 1965 he has been Professor of Physics at University College, Cardiff and is also Professor of Experimental Physics at the Royal Institution of Great Britain. He has devoted much of his research career to the development of visual analogue approaches to the interpretation of X-ray and electron diffraction patterns, and in addition takes a lively interest in physics education in general.

The Schoolmaster:

G. E. FOXCROFT, educated at Trinity College, Cambridge, is Senior Science Master at Rugby School. He has been associated with the Nuffield O level Physics Project and was a team member for the Nuffield A level Physics Project.

IMAGES

A unified view of diffraction and image formation with all kinds of radiation

CHARLES A. TAYLOR
University College, Cardiff

WYKEHAM PUBLICATIONS (LONDON) LTD
(A member of the Taylor & Francis Group)
LONDON and BASINGSTOKE 1978

First published 1978 by Wykeham Publications (London) Ltd.

ISBN 0 85109 620 4 (paper)
ISBN 0 85109 630 1 (cloth)

Printed in Great Britain by Taylor & Francis (Printers) Ltd.,
Rankine Road, Basingstoke, Hants RG24 0PR.

Distribution and Representation

UNITED KINGDOM, EUROPE AND AFRICA
Chapman & Hall Ltd. (a member of Associated Book Publishers Ltd.),
11 North Way, Andover, Hampshire SP10 5BE.

UNITED STATES OF AMERICA AND CANADA
Crane, Russak & Company, Inc.
347 Madison Avenue, New York, N.Y. 10017, U.S.A.

AUSTRALIA, NEW ZEALAND AND FAR EAST
Australia and New Zealand Book Co. Pty. Ltd.,
P.O. Box 459, Brookvale, N.S.W. 2100.

JAPAN
Kinokuniya Book-Store Co. Ltd.,
17–7 Shinjuku 3 Chome, Shinjuku-ku, Tokyo 160–91, Japan.

INDIA, BANGLADESH, SRI LANKA AND BURMA
Arnold-Heinemann Publishers (India) Pvt. Ltd.,
AB-9, First Floor, Safdarjang Enclave, New Delhi 110016.

GREECE, TURKEY, THE MIDDLE EAST (EXCLUDING ISRAEL)
Anthony Rudkin, The Old School, Speen, Aylesbury,
Buckinghamshire HP17 0SL.

ALL OTHER TERRITORIES
Taylor & Francis Ltd., 10–14 Macklin Street, London WC2B 5NF.

Contents

Preface

Two main scientific themes have been at the centre of my research and, by an interesting—but not altogether surprising—coincidence there are strong mathematical links between them. The topics are the diffraction of electromagnetic waves and the physics of musical instruments, and the mathematical concept that connects them is that of Fourier transformation. I am not, however, the kind of physicist for whom the mathematical equations immediately evoke the underlying physics. I accept that mathematics is vital in developing and establishing a theory—but I hold the view that, once established, it ought to be possible to present it in a meaningful way without high-level mathematics. This view, together with the notion that science is a good deal more unified than our traditional fragmented teaching methods sometimes imply, are two philosophical ideas that seem to pervade my thinking and I feel sure that they have coloured my my approach to this book.

My theme is the unity of all diffraction and imaging processes, and I have tried to present what is, in effect, a study of certain practical aspects of the Fourier transform concept as it is applied in modern optics, with practically no mathematics. Here and there I have included a section, marked with an asterisk, that gives a little background mathematics but this is merely for the benefit of those who seek it; to omit these sections should not disturb the continuity.

The book is intended—as are all books in this series—primarily for sixth-form students. I hope, however, that it may find a place as a background reader in physics courses at University and Polytechnic level. I hope too that students of many other sciences at different levels who need to use image-forming systems or diffraction techniques without necessarily going through the discipline of a physics course will find something of interest here. In particular, it may help in understanding the potential and the limitations of the various techniques described.

In writing the book, I have drawn freely on my memory of books read, lectures heard, and demonstrations seen; it is all too easy to remember the substance and to forget the source. I am grateful to many people mentioned specifically but I am also greatly indebted—and apologetic—to those whose contributions are not identified.

Professor Henry Lipson undoubtedly provided the inspiration for my approach to the subject and I owe a great deal to his interest and guidance.

I am grateful to Mr. Noakes, general editor of this series, for planting the idea of writing on this theme. I was fortunate in being invited to lecture to schools at the Royal Institution on this topic and the search for demonstration material for that lecture yielded the basic collection round which the illustrations to this book have been produced. My thanks therefore go to Sir George Porter and Professor Ronald King for providing that opportunity.

Mr. Geoffrey Foxcroft has been my collaborating schoolmaster and has not only fulfilled the scheduled function of making sure that the academic member of the partnership did not take off into the clouds but has also commented, criticized and contradicted in a most rewarding and helpful way. It is a great comfort to know that he has read the manuscript with a most school-masterly eye for detail as well as for the broad issues. Nevertheless, any errors and heresies that undoubtedly will be discovered must remain my sole responsibility.

Bob Watkins, the photographer in my department, has played a very significant part in producing a high proportion of the illustrations and my colleagues George Harburn, Alan Fowler and Mike Evans have also given valuable help in preparing illustrations. My thanks go to all these people as well as to those who have agreed to let me use their illustrations (acknowledgments of which are given elsewhere), to Mr. W. Sutherland of Velindre Hospital and Professor Evans of the Welsh National School of Medicine for providing illustrations—some of which were specially produced, to Jayne Tonks, an electronics technician in the department who allowed herself to be photographed and to various colleagues in my department who have helped me to clarify various ideas in discussion; Dr. Barrie Jones and his colleagues at the Open University have drawn me into helpful discussions of a course that they have prepared on a closely related theme; my secretary Mrs. Valerie Chown deciphered my manuscript in her usual helpful way; and finally my gratitude goes to my wife and family for their patient forbearance and encouragement.

CHARLES TAYLOR

1. The physics involved in forming an image

> " However entrancing it is to wander unchecked through a garden of bright images, are we not enticing your mind from another subject of almost equal importance? "
>
> Ernest Bramah
> *Kai Lung's Golden Hours*

1.1. *What are the common features of all image-forming systems?*

We all possess image-forming devices as part of our anatomy, and a study of the precise way in which our eyes operate will give us a good basis for starting our discussions. Many textbooks—at all levels—have been written which include detailed accounts of the physiological structure of the eye, or of the paths traced by the rays of light through the lens and the various media with which the eye is filled. I do not propose to follow this pattern at all; we shall not consider detailed operation but rather the broad essential functions that must be performed in order to see.

Imagine a lecturer standing in front of his audience. He is clearly gaining information continuously through his eyes. One student has a rather florid tie; a woman is examining the state of her make-up in a mirror; a man is having trouble with his fountain pen; two people in the back row have dozed off to sleep, and so on. Similarly the audience receives information about the lecturer. This information is, of course, conveyed from one to the other via the light in the room. That is a very obvious statement since, if we put the room in total darkness, the transfer of information ceases immediately. (In fact to put the room in total darkness is suggested as the first experiment which dramatically draws attention to the first part of the image-forming process.) What happens normally is that the light in the room falls on the various objects in it and is both scattered and modified; as we shall see later, the scattering depends on the shape and texture, the colour may be modified by selective absorption of different wavelengths, and there may also be changes in polarization. It is the scattering process, however, that will be of most concern to us in this book.

The word ‘ scatter ’ needs considerable elaboration and indeed, a large part of Chapter 2 is devoted to it. For the moment it will suffice if we interpret it as meaning that the waves bounce off the object in some way and hence the original pattern of waves is disturbed by the presence of the object.

The next point to recognize is that only a tiny fraction of this scattered light ever reaches the eye of the observer but, nevertheless, his eye and brain together are able to abstract an extraordinary amount of information from it. That all the information is present in the scattered light may easily be demonstrated by a second ‘ silly ’ experiment: a small card is placed in front of the eye. Obviously the information transfer ceases since the waves are prevented from entering the eye; the waves that carried the information must therefore be falling on the card. But, if the card were replaced by a photographic plate, the resulting record would clearly not be useful; it would just be more or less uniform blackness. Nevertheless, all the information must be potentially available at that point and later we shall need to consider how it could be extracted and in what form it exists.

A third, and probably the most useful, of our introductory ‘ silly ’ experiments which drives this point home is to place a rather improbable slide (I often use a cartoon or a very ancient poster—in fact anything that the audience is unlikely to expect) in a projector without the lens in position. The patch produced on the screen is then completely uninformative and yet a moment's thought makes it clear that all the necessary information for the production of a complete image of the slide must be present on the screen, though in a form which is not directly interpretable. Fortunately in this particular example all that is necessary to interpret the pattern is to replace the lens in the projector. But it should be pointed out that the lens cannot possibly ‘ know ’ what is on the slide; all it can do is to rearrange the light that has been scattered during its transit through the slide. Once again therefore we see quite clearly that the information about the slide must have been on the screen in some form or other all the time.

Technically the pattern on the screen with no lens in the projector is called a hologram—though this word is reserved by some authors for cases in which the illumination is rather special (see section 4.3). The term simply means that *each* point on the screen or on a photographic plate is receiving information from *every* point on the object (see Fig. 1.1 (*a*)).

We can demonstrate experimentally that this is so with a simple slide—a good subject would be a thick black cross—still with no lens in the projector. Place the projector as far away from the screen as is possible and then use a converging lens of longish focal length (say about 500 mm, though a shorter one may be necessary if the projector–screen distance is not large) to form a reduced image of the slide on the screen. You will find that a reduced image of the whole slide can be produced with the lens at any position within the patch of light (fig. 1.1 (*b*)). This reinforces the assertion that all points on the screen contain information about all points on the slide.

Thus, the first broad stage in the process of image formation may be described as the scattering of the incident radiation by the object. It could be regarded as a process of coding, and, in order to interpret the results, we need a decoding process which in the simple systems we have so far considered merely involves the lens.

I must now digress slightly to draw attention to an important alternative way of seeing. When the lights were out in my first 'silly' experiment, one piece of information that the lecturer might have picked up, even in the dark, could have been that someone was smoking: in other words he might have seen the glow of a cigarette. Here the object is self-luminous and is producing its own radiation. We shall see later however that the subsequent process of forming an image from the *radiation* pattern of the self-luminous object is almost identical with that of forming an image from the scattering pattern of a non-luminous object. (It is in fact *exactly* the same if the light being scattered is incoherent but the process is significantly different from the scattering of coherent light.)

What does the lens need to do? Fig. 1.1 (*a*) can help in the explanation; the lens must take all the elements of information which pertain to, say, point A on the slide which would otherwise have been dispersed to points P, Q, etc. on the screen and put them together at a single point on the screen which will then become the image of point A. This is a very remarkable operation indeed but, because it is so familiar, and because we can draw neat ray diagrams showing how it occurs, we tend to take it for granted. Unfortunately the simple ray diagrams of geometrical optics hide a great deal of the complexity of the operation and, while telling us where and how big the image is, do not, for example, draw our attention to the important problems of contrast, brightness and sharpness of the image.

Thus the second stage of image formation is the recombination of the scattered or radiated waves to form an image. Later we shall be considering nomenclature in more detail but perhaps it is appropriate

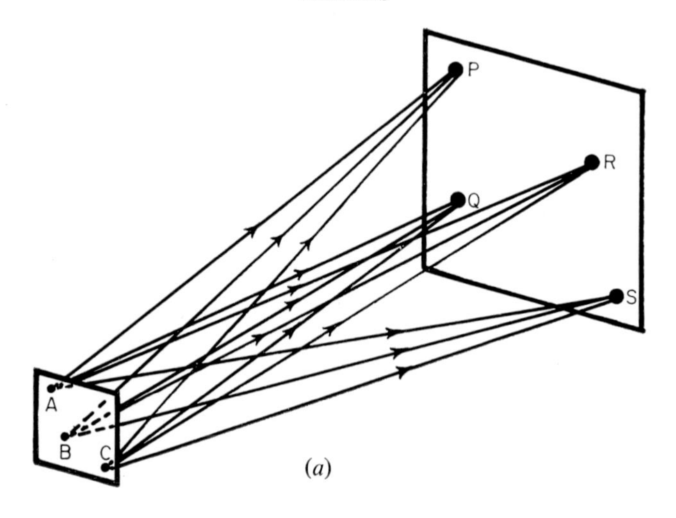

(*a*)

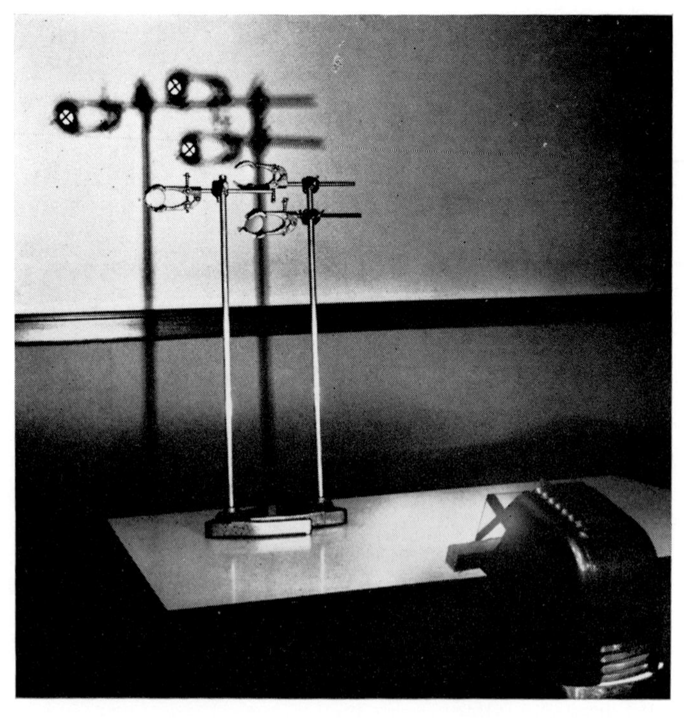

(*b*)

Fig. 1.1. (*a*) The hologram relationship: each point of the slide contributes scattered light to each point of the screen. (*b*) The hologram relationship: wherever the lens is placed a *complete* image of the slide is produced.

to comment at this point that topics such as X-ray, electron or neutron diffraction will come within the general purview of our discussions just as much as the scattering of light, radio-waves, etc.

We have thus reduced the image-forming process to two portions, scattering, or radiation, and recombination—but we have said nothing specific about the operation known as focusing. It is a very familiar one; most people can perform it almost instinctively. It is nevertheless worth considerable attention. When one focuses the image of the slide on the screen what is the real essence of the operation? A few moments' consideration will lead to the conclusion that all we are really doing is to make the image look as nearly as possible like we think it should look! If for example, the slide has printing or lines on it, we assume that they have sharp edges (fig. 1.2 (*a*)); a slide deliberately prepared out-of-focus (fig. 1.2 (*b*)), or a piece of fluted glass which has no sharp definition, provides a much more difficult problem for the operator (fig. 1.2 (*c*)–(*f*)). In fact there are only two possible ways of focusing; we may measure the various distances involved and calculate where to put the lens in relation to the other components; or we may, by trial and error, make the image resemble as closely as possible our notion of the object. It is important to realize that these are the *only* alternatives and we shall see later how they influence various applications.

One practical way of trying to surmount the problem is to focus on some part of the object that we recognize and then assume that the

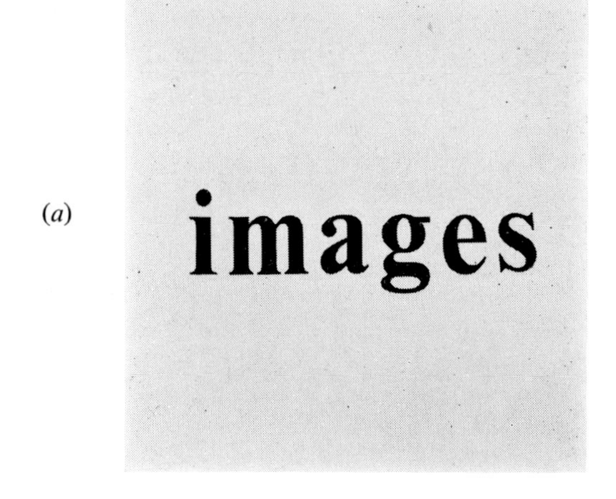

Fig. 1.2. (*cont. overleaf*).

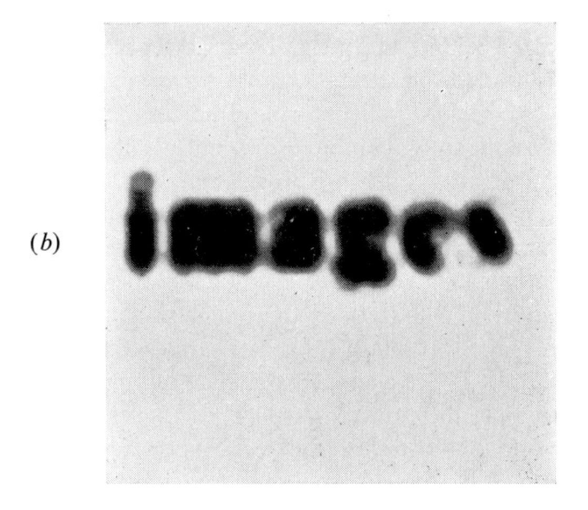

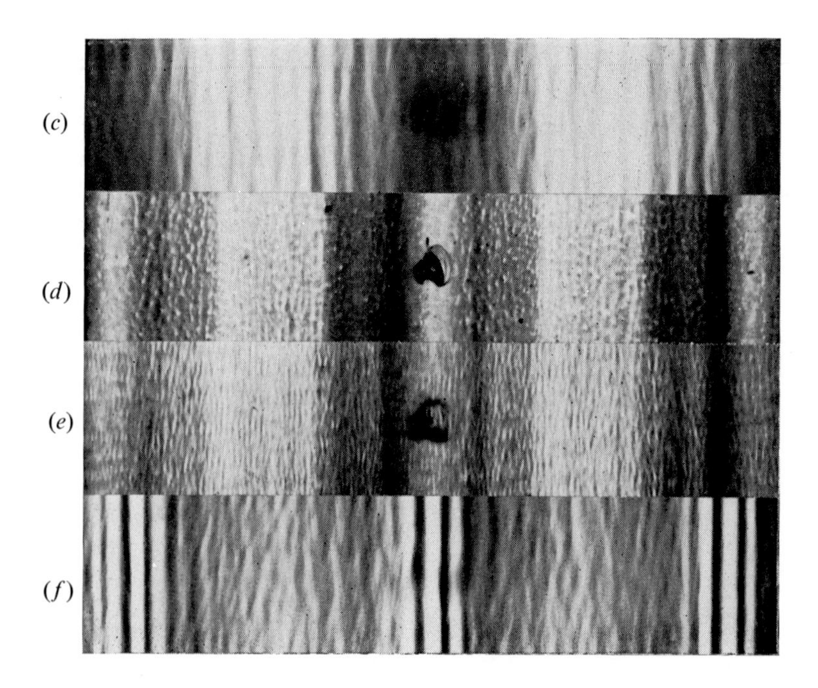

Fig. 1.2. (*a*) Object with sharp edges in focus. (*b*) Object with sharp edges out of focus. (*c*)–(*f*) Fluted glass with four different positions of focusing lens; in (*d*) the chip is in focus.

rest must then be in focus—and this is the basis of one of the powerful techniques in X-ray diffraction (pp. 96–99). Fig. 1.2 (*d*) illustrates this point. Focusing is an operation which is always taken very much for granted; it is in reality of such great significance in understanding many of the processes of image formation that we shall need to consider it later in much more detail. We can summarize the argument so far by saying that the process of seeing or imaging involves two basic elements—scattering or radiation, and recombination—with the vital operation of focusing linking the two. We should now consider the limitations that arise before developing the ideas in greater detail.

1.2. *What are the limitations which restrict the precision of an image?*

In order to investigate the principal sources of limitation in image formation it is convenient to continue our consideration of the experiment in which the slide was placed in the projector without a lens. But the question we now ask is " In what form is the information present on the screen? ". The amplitude, or intensity, is more or less uniform across the screen and hence the only other variable possible would seem to be that of phase. In this example, however, the light consists of many separate little trains of waves originating from different parts of the source and having many different wavelengths. For any one of these trains there will be phase relationships arising as a result of the scattering by the object and these lead to some kind of interference pattern on the screen. Because the trains are arriving randomly, both in terms of their wavelength and of their direction, all the interference patterns will be different and the change from one to another will occur at such a high rate that we see no effect and the screen appears uniformly bright. (This is a manifestation of the phenomenon that we call ' incoherence ' which will be discussed at greater length in Chapter 2. Under coherent conditions it is possible to record the phase pattern and the result is the technique known as holography which is described in Chapters 4 and 6.)

Let us first consider the purely geometrical problems of the introduction of phase relationships during the scattering process. Here I shall simplify matters by considering the effect of just one of the wave trains—in this case a plane wave—arriving simultaneously at two particular points on the object. A simple experiment with striped string or with the Nuffield O-level wave form apparatus is a convenient introduction (see fig. 1.3). The string is supposed to represent the paths of waves, scattered from two points on an object, arriving at the screen, and the stripes—which indicate schematically the crests and troughs—help us to see what relative changes in phase will occur.

In fig. 1.3 (*a*) the scattering points are a long way apart and we can see that the distance on the screen between points at which the two scattered waves are in phase or are in opposite phase (i.e. 180° out of phase) is relatively small; in other words there is a pretty rapid alternation of phase difference as we move the point of observation across the screen. If the points are closer together as in (*b*) we have to move further for a corresponding change in relative phase and, if the points are very close together as in (*c*) we have to move a very considerable distance before any significant phase difference is produced at all. This kind of experiment allows two points to be made quite clearly.

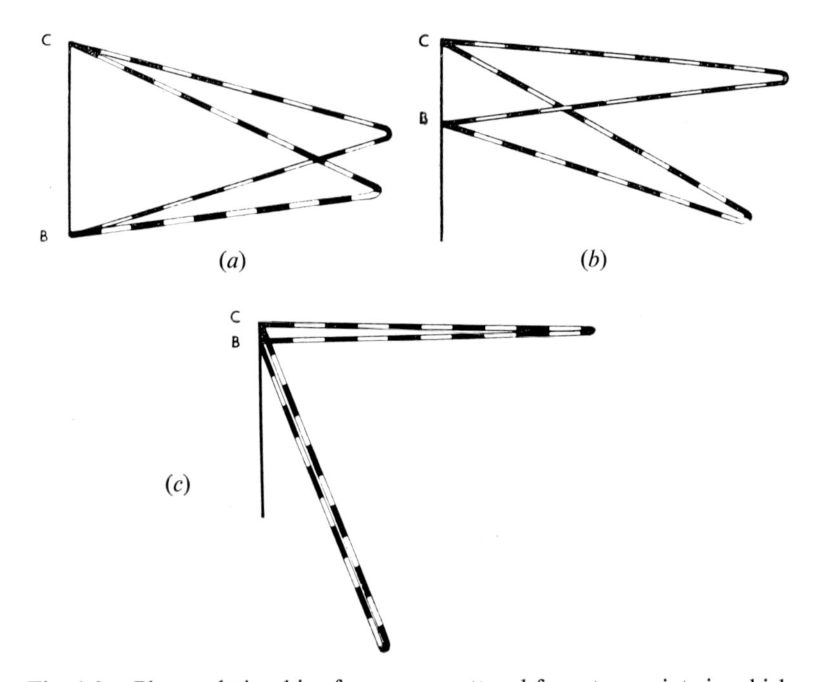

Fig. 1.3. Phase relationships for waves scattered from two points in which the black stripes indicate crests and the white stripes troughs. (*a*) Scattering points far apart. (*b*) Scattering points closer together. (*c*) Scattering points less than one wavelength apart. From *The Physics of Musical Sounds*, by C. A. Taylor, by permission of Hodder & Stoughton Educational.

First, unless the object is large compared with the wavelength, scattering over a fairly wide angular range is necessary in order to obtain significant changes in relative phase and hence significant coding of information. Secondly, if the object is smaller than the wavelength (roughly), no significant relative phase change can be made at all, and hence it will be impossible to form an image which will contain any useful information by any system. This is a simple approach to the idea of limit of resolution but is quite valid.

Thus the first limitation on the precision of the image is imposed by the relationship between the wavelength of the radiation and the size of detail to be recorded. The second limitation is quite closely related and can be illustrated by the same experiment. If scattering occurs over a wide range of angles, it may well be that much of the scattered radiation will miss the recombination system. The effect of taking in only a small cone of radiation with little relative phase change across it will be much the same as that of using too large a wavelength. Thus the aperture of the recombining system is the second limitation.

The third fundamental limitation depends on the nature of the scattering process itself. We shall take just three examples. If we are using visible light it will be difficult to image transparent objects since the interaction between the object and the incident radiation will not be very great; with X-rays it will be impossible to record any information about the nucleus of an atom because the X-rays are all scattered by the electrons and do not interact with the nucleus; with microwaves quite different results occur if the scatterers are made of dielectric rather than of conducting material even if the objects have the same size and shape.

A fourth fundamental limitation concerns the nature of the intervening medium. For example an optical telescope is no use in a fog; radio waves of certain frequencies cannot penetrate the ionized layers in the upper atmosphere; ultrasonic radiation is damped out very quickly indeed in water; ultraviolet radiation will not pass through glass whereas visible light will; non-uniformities in the medium will introduce distortions in the image, e.g. the 'shimmer' of distant objects on a hot day.

Limitations of a different kind arise from the way in which the radiation is detected. For visible radiation we may use our eyes, photographic film, or electronic apparatus such as television cameras and other photo-sensitive devices. In all of these examples the response is proportional to the intensity of the radiation, which is proportional to the square of the amplitude, and no direct record of relative phase is made. All these detectors are capable of distinguishing the effects of differing wavelength, though of course the relationship between the colour we see or record photographically and the actual wavelength present is not a simple one and involves some interesting problems of psychology and physiology which are beyond the scope of this book.

Waves in the radio region, however, will not affect photographic plates, and we need to use dipoles or some other kind of radio antennae. These devices respond directly to both the amplitude and the phase of the signal, and so we are able to record the patterns in greater detail. The contrast between the detection of the shorter wavelength electromagnetic radiations and of waves in the radio region is an interesting one: for the former we can record the details of a whole scene simultaneously but only in terms of the square of the amplitude; for the latter, on the other hand, we can record both the amplitude and the phase but only at one point for each detector at a given moment.

The main limitation imposed by all detectors is that of sensitivity. There was a time when the eye was the most sensitive detector of visible radiation, but it is now possible to do better with both photographic and electronic devices.

X-rays, ultraviolet rays and γ-rays can be detected by the same methods as visible light (except, of course, for the eye) and hence no phase information can be recorded; ultrasonic radiation and sound waves are detected by microphones which are really instantaneous pressure measurers and, since they can distinguish between pressures that are higher or lower than atmospheric, phase information can be recorded. Again, as with radio waves, each detector can record information only at one point at a time.

1.3. *To what kinds of radiation do these principles apply?*

The same general principles apply to any kind of radiation though the physical mechanism of the scattering processes may be entirely different. The techniques used to recombine the image may also be totally dissimilar. It is important, however, to be able to grasp the interconnections between these apparently unrelated techniques, as we shall see later.

Fig. 1.4 is a chart which gives the wavelengths corresponding to a great many different radiations and a list of common objects of sizes comparable with the various wavelengths is also included. Since we saw that it is virtually impossible to impress on a wave any information about detail less in size than its wavelength, we can see that, in order to image an object of the size suggested in the table, a radiation which appears lower in any column than the object must be selected.

In the electromagnetic spectrum the wavelengths range from thousands of metres down to fractions of a nanometre (1 nm$=10^{-9}$ m) but the nature of the physical interactions with objects changes considerably. Radio waves are scattered by both conductors and non-conductors, though in somewhat different ways; radar systems may

Typical objects in the size range	Order of magnitude in metres	Frequency in hertz of sound waves having this wavelength	Frequency in hertz of electromagnetic waves having this wavelength	Accelerating voltage for particles having this equivalent wavelength	Energy of uncharged particles having this equivalent wavelength
City	10^4		3×10^4		
Airfield	10^3		3×10^5		
Jumbo jet	10^2	$3{\cdot}3$ (Infra sound)	3×10^6		
House	10	$3{\cdot}3 \times 10$	3×10^7 (Radio)		
Child	1	$3{\cdot}3 \times 10^2$ (Audio sound)	3×10^8		
Sparrow	10^{-1}	$3{\cdot}3 \times 10^3$	3×10^9 (Infra red)		
Bee	10^{-2}	$3{\cdot}3 \times 10^4$ (Ultra sonics)	3×10^{10}		
Daphnia	10^{-3}	$3{\cdot}3 \times 10^5$	3×10^{11}		
Amoeba	10^{-4}	$3{\cdot}3 \times 10^6$	3×10^{12}		
Chlorella	10^{-5}	$3{\cdot}3 \times 10^7$ (Acoustic surface waves)	3×10^{13} (Visible light)		
Typhoid bacillus	10^{-6}	$3{\cdot}3 \times 10^8$	3×10^{14} (Ultra violet)		
Smallpox virus	10^{-7}	$3{\cdot}3 \times 10^9$	3×10^{15}		
Turnip yellow virus	10^{-8}	$3{\cdot}3 \times 10^{10}$	3×10^{16}		
Vitamin A molecule	10^{-9}		3×10^{17} (X-rays)		
Carbon atom	10^{-10}		3×10^{18}	Electrons $1{\cdot}5 \times 10^2$	Neutrons of thermal energies
	10^{-11}		3×10^{19} (γ-rays)	Electrons $1{\cdot}5 \times 10^4$	
	10^{-12}		3×10^{20}	Electrons $1{\cdot}5 \times 10^6$	
	10^{-13}		3×10^{21}	Protons 8×10^4	
Uranium nucleus	10^{-14}		3×10^{22}	Protons 8×10^6	

Fig. 1.4. Orders of magnitude of various objects together with the frequencies or energies of waves having wavelengths of similar dimensions.

use either longer radio waves to give general long-range warnings or very short waves to give quite detailed images. Infrared radiation can pass through rock crystal but is scattered and absorbed by glass. The visible region is the most familiar and its properties need no elaboration at this stage. X-rays and γ-rays interact quite differently and are scattered largely by the electron clouds surrounding each atom in a solid.

Acoustic waves follow yet a further different pattern. Scattering for all the longer wavelengths, including all the audio region, is almost purely mechanical and depends on the geometry and surface properties of the object. At the higher frequencies of ultrasonics much more penetration occurs and the interactions are more complex. One can make practical use of the refractive index between two media for ultrasonics (the ratio of the velocities of sound in the two) since reflections occur from boundary layers between media of different refractive index.

Finally various particles such as electrons, protons or neutrons may be used for imaging and each will have its own peculiarities quite apart from the resolution which depends on its equivalent wavelength. Electrons—in the electron microscope—have been very powerful explorers of objects beyond the limits of optical microscopes, but it is necessary to learn how to interpret the results: it is no longer possible merely to look at the resulting picture in a subjective way. Neutrons can be used and because of their lack of charge may penetrate through the electron shells to the nuclei. We shall see in the final chapter how the use of these various radiations works out in practice.

2. A closer look at radiation, scattering and diffraction

> "We all *know* what light is; but it is not easy to *tell* what it is."
>
> Samuel Johnson
> Boswell's *Life of Johnson*
> Vol. iii, 12 April 1776

2.1. *Radiation and the idea of coherence*

All imaging processes depend on the interpretation either of radiation patterns from sources or of scattering patterns of radiation interacting with an object. In both cases the radiation is the vital element and we must start our detailed study by looking closely at some of the ways in which radiation behaves.

We shall be talking chiefly about the wave aspects of radiation and so it may help to create the right pictures in our minds if we think first of waves on the surface of water in a ripple tank. Suppose we use a single dipper with a fairly small point to create the waves, and suppose further that we drive it sinusoidally so that the point is rising up and down at a completely regular frequency for as long a time as we wish. The result is an infinite train of circular waves radiating out from the point. Fig. 2.1 shows this diagrammatically and also gives a photograph of actual waves.

If we choose two points such as A and B which are equidistant from the source, the wave is always at the same point of its cycle for each; cork floats placed at A and B ride up and down exactly in step with each other. The same is true if we choose any other point on the same circle as A and B. This relationship is an aspect of the wave phenomenon that we call *coherence*. In this example in which we relate the oscillations at a fixed distance from the source (and, since the waves travel at constant speed, this implies that we are considering the behaviour of points at identical time intervals from the commencement of the oscillation) we are dealing with *spatial* coherence. It could be described as the study of phase relationships between the displacements at different points in *space* at a particular instant of time.

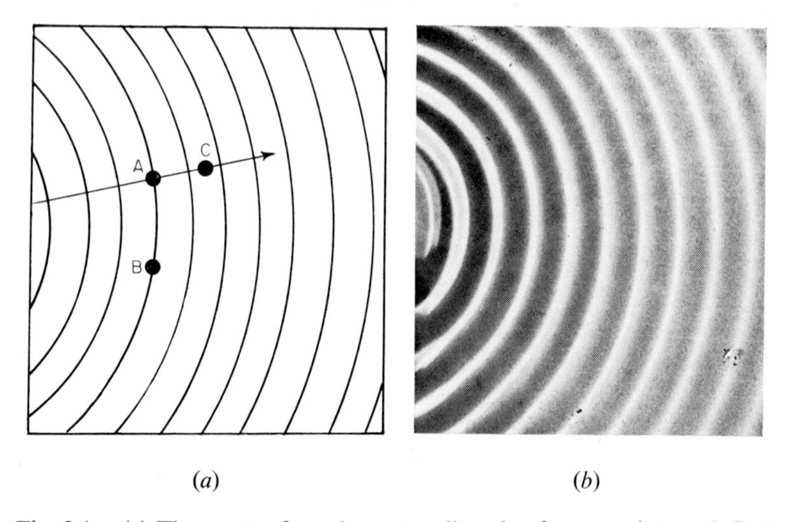

Fig. 2.1. (*a*) The crests of regular waves diverging from a point: cork floats at A and B are always in step and demonstrate spatial coherence; floats at A and C follow each other at the same frequency with constant phase difference and illustrate temporal coherence. (*b*) Photograph of corresponding waves on a ripple tank.

If, on the other hand, we concentrate on a single point such as A and consider the oscillation as a function of time, we shall find, in this example, that a cork float placed here rises up and down in a perfectly regular way as time goes on; there are no discontinuities or irregularities in the graph of its displacement against time. This is an aspect of wave behaviour that we call time or temporal coherence. Since the wave is travelling at a constant velocity in this example, we can use the expression

$$distance = velocity \times time$$

to substitute distance for time, provided that the distance is measured along the direction of travel of the wave. Thus the behaviour of a cork float at C follows the behaviour of A with a constant time delay. We could thus explore the behaviour of the displacement at A as a function of time by studying the behaviour of the displacement at all points along a line such as AC at a particular moment of time. This trick is valuable experimentally and we shall use it quite frequently, but there are complications in certain cases which are not as straightforward as the arrangement used here. For this reason I prefer to describe *temporal coherence* as the study of phase relationships between displacements at different instants of *time* at a particular point in space. This also has the advantage of preserving a certain symmetry with our earlier description of spatial coherence.

Suppose now that, instead of using a regular sinusoidal drive on the dipper, we operate it in some random way. Circular waves still spread out from the dipping point and they still travel at a constant velocity. The displacements at points A and B are still in step with each other but the graph with time of the displacement at A is no longer a regular sine curve. Thus we could say that the spatial coherence remains unimpaired but the temporal coherence has disappeared (fig. 2.2).

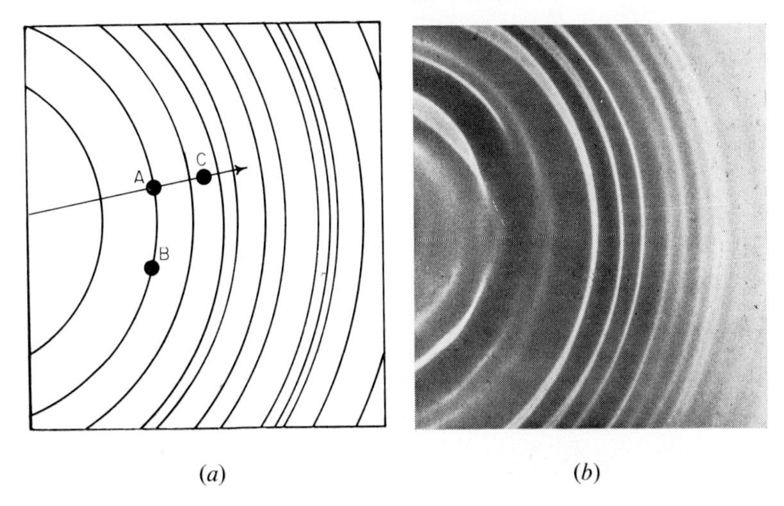

(*a*) (*b*)

Fig. 2.2. (*a*) Irregular waves diverging from a point: A and B still exhibit spatial coherence but A and C no longer exhibit temporal coherence (*b*) Photograph of corresponding waves on a ripple tank.

Now suppose we introduce a number of different dippers randomly positioned, and all driven randomly (fig. 2.3). It is fairly clear that both the temporal and spatial coherence will disappear and we end up with ' incoherent ' waves. Is it a clear-cut ' black-or-white ' situation? In other words must the radiation represented by the ripples be either coherent or incoherent, or is there a state in between?

Consider the temporal coherence first. Suppose, in the example of fig. 2.1, that instead of driving the dipper with a pure sinusoidal waveform we feed two frequencies in simultaneously. The combined result is the well known ' beat ' effect. If the two frequencies are very close together the beat ' envelope ' is very long and so there are longish periods when A oscillates regularly. If we again make use of the time-distance transformation this means that we can move a fair distance

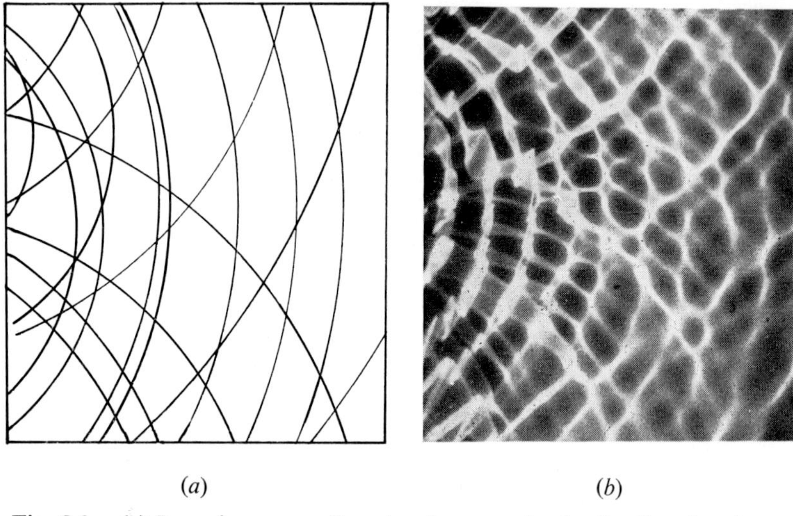

(*a*) (*b*)

Fig. 2.3. (*a*) Irregular waves diverging from randomly distributed points: both temporal and spatial coherence have disappeared. (*b*) Photograph of corresponding waves on a ripple tank.

towards C from A without a discontinuity in the oscillation. If however the frequencies are further apart, the trains of continuous waves will be much shorter! It turns out that the more indefinite the wavelength (or frequency) the shorter is the train and for practical purposes we often describe this quantitative aspect of the temporal coherence as the 'coherence length'. Fig. 2.4 illustrates these principles.

Waves, of whatever nature, that have well defined wavelength, are often described as monochromatic, though strictly this term should apply only to light. Highly monochromatic waves have a long coherence length and are very coherent temporally; waves with a broad frequency distribution have correspondingly shorter coherence lengths and are much less coherent. We have thus introduced the ideal of *partial* temporal coherence.

Suppose now we go back to fig. 2.3 and add a small aperture through which the waves must pass (fig. 2.5). We now find that there *are* phase relationships over a certain region in space. In general the *smaller* the aperture the larger the regions of phase correlation. In other words the smaller the aperture the greater the degree of spatial coherence. So here again we have an intermediate effect—partial spatial coherence. Notice, however, in the actual ripple tank photograph of fig. 2.5 (*b*) the *amplitude* of the waves emerging from the slit

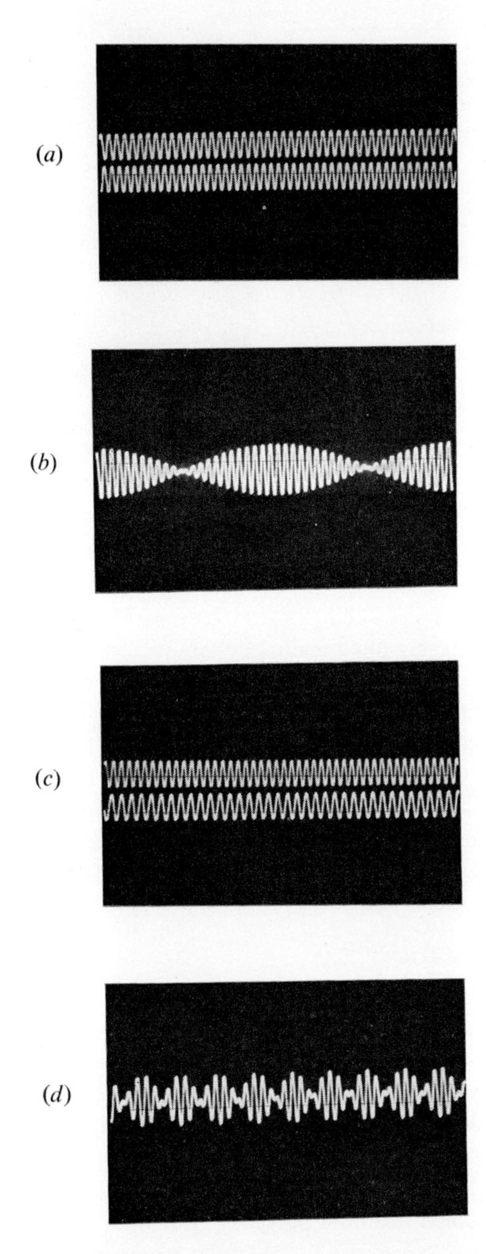

Fig. 2.4. (*a*) Two sine waves of almost the same frequency. (*b*) The sum of the waves in (*a*): the coherence length is relatively long. (*c*) Two sine waves with a much larger frequency difference. (*d*) The sum of the waves in (*c*): the coherence length is much shorter than for (*b*).

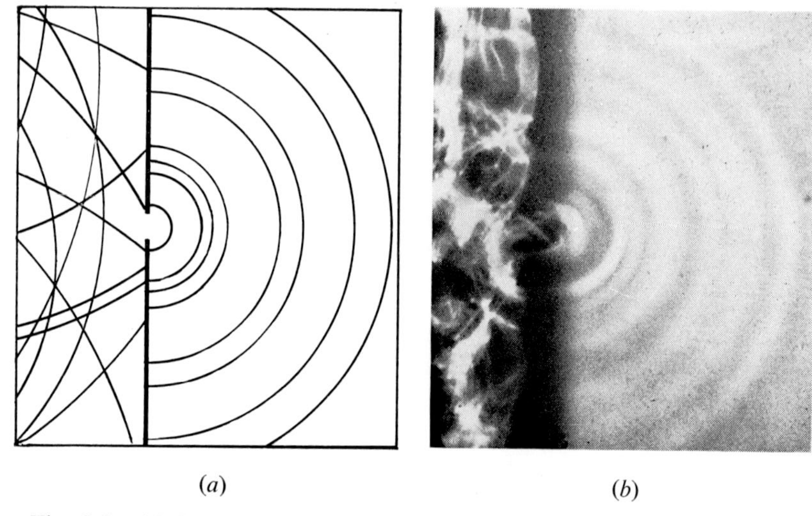

Fig. 2.5. (*a*) Restoration of spatial coherence by interposition of narrow slit. (*b*) Photograph of corresponding waves on a ripple tank.

is very small; we have gained spatial coherence only at the expense of a large reduction in amplitude. As we shall see later, this is a point of some importance.

This has been an entirely non-mathematical way of looking at partial coherence but it will serve our purpose quite well. The mathematical treatment is considerably beyond the level of this book and only becomes important when specific systems have to be designed. We shall however return to the topic in a little more detail at the end of this chapter.

2.2. *Coherence in practice*

Now let us see how the ideas of coherence apply and consider their implications for the various forms of radiation that can be used in imaging.

First let us think about sound waves. A single musical instrument, such as a clarinet, will produce relatively coherent waves both temporally and spatially. A single note of definite pitch will have a coherence length that is simply equal to the velocity of sound multiplied by the time for which the note persists. But how can we attempt to measure the spatial coherence? With the water waves we imagined two floats and considered whether they were remaining in a constant phase relationship. This can be observed *directly* in the case of the water waves and could also be observed relatively directly for sound;

microphones could be placed at the relevant positions and the resultant oscillograph traces compared. Direct observation is not so easy with light and other higher frequency radiations however. An almost universal technique is to find out whether interference patterns can be observed and the extent to which this is possible gives a measure of the degree of coherence. We shall consider this in greater detail in section 2.8, but for the moment let us see how the idea works out with simple experiments in sound.

If an instrument is played in a room, the reflections from the walls will soon build up an interference pattern which will result in the sound varying in loudness from point to point. This effect is easily verified if you have not met it before by moving your head slightly from side to side in a room in which a high-pitched whistle is sounding. As the head moves even a small amount the sound changes in loudness. Suppose however that an instrument is played with vibrato. This usually means that the frequency is ' wobbling ' slightly; this in turn means an uncertainty in the frequency and a reduction of the coherence length. The result is that the interference pattern is either destroyed or moves about and so one ceases to be able to distinguish it clearly. A small source of random noise—such as steam escaping from a kettle—would have an extremely short coherence length but could still be said to create waves with spatial coherence. A very widely distributed source such as an audience clapping would produce little coherence either temporally or spatially.

Ultrasonic sources are nearly always highly coherent both temporally and spatially.

Now let us turn to the electromagnetic spectrum. Radio waves at all frequencies are usually highly coherent if produced by a single transmitter and aerial system. By its very nature, radio-frequency radiation stems from oscillations maintained in electrical circuits and these tend to be continuous for long periods of time.

One of the simplest demonstrations of the coherence of radio frequency radiation is the interference effects that occur when an aeroplane flies over and a television picture alternately fades and brightens. In this example the direct waves from the transmitter and those reflected from the plane arrive together at the receiver aerial and *because* the radiation is coherent they will have a specific phase relationship for a given position of the aeroplane. If the waves arrive in step the picture will brighten and if the waves are exactly out of step it will fade. Then, as the aeroplane moves, the path difference between the two sets of waves changes and so the phase difference changes to give the familiar alternating effect.

Now let us move up through the frequency range of the electromagnetic spectrum. In the infrared region, and indeed through all the higher ranges, the processes that generate the radiation are all concerned with separate particles in some way. Atoms change from one energy state to another and in so doing emit a quantum of radiation —a photon as it is usually called. When an electron that has been accelerated under the influence of a high potential collides with a target X-rays are emitted—but each electron colliding initiates a little packet of waves separately and independently of the others. A nucleus decays and emits a gamma ray—but again each nucleus in a mass of radioactive material behaves more or less independently of all the other nuclei. The result is that the radiation emitted lacks both temporal and spatial coherence and resembles the water waves of fig. 2.3.

This does not mean, of course, that one cannot have coherent radiation in the higher frequency ranges of the electromagnetic spectrum. Before 1960, the only way of producing coherent radiation in this range was to use the trick which we mentioned in the discussion of water waves; if the radiation is sent through an aperture of some kind then some degree of spatial coherence will be imposed, and if it is put through some kind of filter or monochromator then, the narrower the frequency pass band of the filter, the greater will be the temporal coherence. The difficulty is that the greater the coherence (of either kind) achieved, the more light is thrown away. Thus experiments with coherent light were not easy to perform. Alternatively, temporal coherence can be achieved if a monochromatic source is used in the first place, e.g. in the visible region a sodium flame or lamp gives light in which the predominant frequency is about 5×10^{14} Hz and has sufficient temporal coherence to make it very useful for interference experiments though its brightness is not very high. In 1960, a laser was made to operate in the visible region and one of its characteristics is that its frequency band may be so narrow that a coherence length of tens of kilometres may result! It would be out of place to go into any great detail of the mechanism here but we may find it helpful to devote half a page or so to the general features of laser radiation which has revolutionized experimental optics.

For optical experiments we find that solid-state lasers are not as generally useful as gas-phase lasers. The ruby laser emits single flashes of immense power which find all kinds of applications in other fields, but the helium–neon or argon-ion gas lasers emit continuous radiation which comes very close to the kind of radiation emitted by a radio transmitter, from the coherence point of view. Crudely speaking, a gas laser consists of a glass tube in which an electrical glow discharge is taking place, so that, because of the collisions occurring in the discharge, a great many of the gaseous atoms are in an excited state. Normally they would revert to the unexcited state randomly and emit quanta of light waves in so doing. In the laser system a pair of strongly (but not completely) reflecting mirrors is placed, one at each end, so that light from within the tube is reflected back and forth many times. The mirrors are specially coated so that they reflect strongly only one of the characteristic spectrum wavelengths from the discharge, and the remarkable effect is that, as the beam travels up and down the tube, it triggers off the excited atoms and they emit their quanta in step with each other. If the laser is carefully designed it can be made to emit a so-called ' uni-phase ' wave front—that is the radiation is not only highly coherent temporally but is also spatially coherent across the whole width of the beam; the resulting radiation emerges in a highly parallel beam without the need for any optical system. The light energy is so coherent that interference effects can be produced by merely placing two slits in the beam without any source slit. Two separate lasers can be made to interfere and produce ' beats ' just as two tuning forks can in the acoustic case. Very strict control of temperatures and voltages is needed to make this trick work because the precise frequency at which the system ' lases ' depends on a variety of factors. The basic frequency of a helium–neon laser is about $4{\cdot}7\times10^{14}$ Hz and it follows that, to produce an audio beat stable to ±10 Hz, the frequency of each laser must be stable to better than 1 part in 10^{13}. High precision indeed!

2.3. *Coherence, bandwidth and the Uncertainty Principle*

Before going much further with this discussion, the somewhat difficult concept of bandwidth will need to be elaborated. I propose to make this a starred section and it could be omitted, certainly at first reading.

* Whenever we use radiation of any kind for transmitting information from one point to another—whether the information is morse code, speech, music, scientific data from space probes, or television pictures—the very fact that information is being sent imposes changes in the nature of the waves. Let us consider the simple example of using radio waves of frequency f_c (the so-called *carrier* frequency) to convey a continuous pure musical tone of frequency f_m (the so-called *modulation* frequency). Fig. 2.6 shows diagrammatically how this is done; (*a*) shows the original carrier wave, (*b*) shows the information to be transmitted made everywhere positive by adding a constant and (*c*) shows (*a*) multiplied by (*b*), the

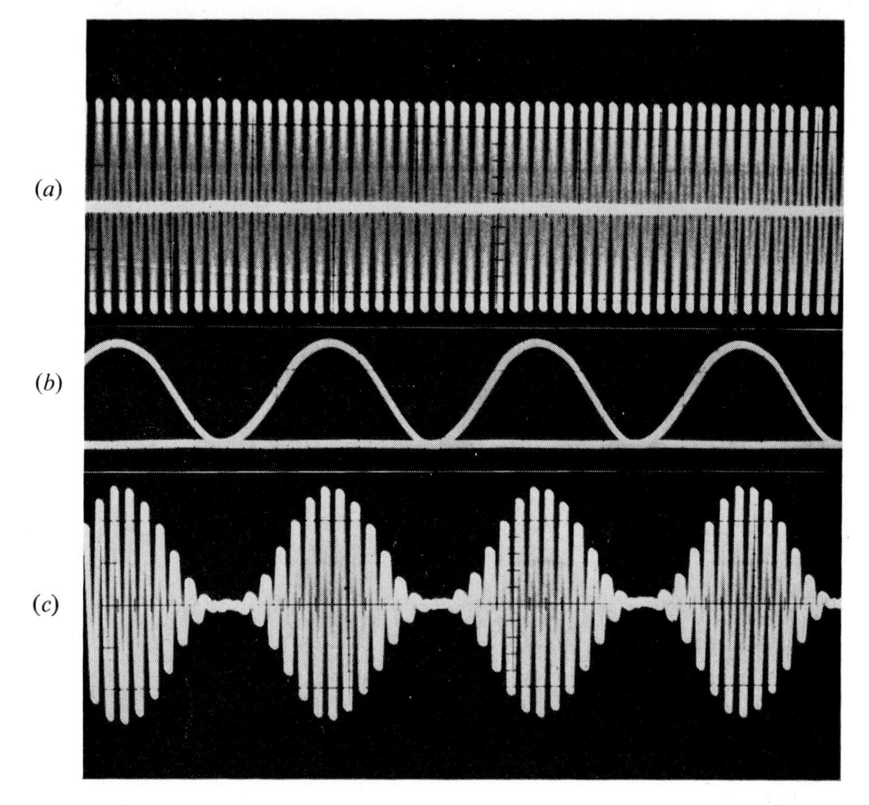

Fig. 2.6. (*a*) Trace of carrier wave. (*b*) Trace of modulating wave. (*c*) Trace of modulated wave.

actual modulated wave which goes from transmitter to receiver. Mathematically we can represent (*a*) by the equation $y_c = A \sin 2\pi f_c t$, and (*b*) by the equation $y_m = B + B \sin 2\pi f_m t$.

The modulated wave (*c*) is then the product

$$y_c \times y_m = [A \sin 2\pi f_c t]\ [B + B \sin 2\pi f_m t],$$

which we can rewrite, using standard trigonometrical formulae, as

$$y_c \times y_m = AB \sin 2\pi f_c t + \tfrac{1}{2}AB \cos 2\pi(f_c - f_m)t - \tfrac{1}{2}AB \cos 2\pi(f_c + f_m)t.$$

The important point to notice is that we now have the two frequencies $(f_c + f_m)$ and $(f_c - f_m)$ present besides the basic carrier frequency f_c. The same argument could be used for any kind of information and if we had many different modulation frequencies—as for example in the transmission of music—many frequencies would be present in the modulated wave. If F_m is the maximum frequency introduced in the modulation, then all these additional frequencies are contained in the band $f_c - F_m$ to $f_c + F_m$ and hence the *bandwidth* required to transmit this information is $2\,F_m$. The consequences of this principle are very wide-ranging. It is, for example, one of the reasons why television pictures have to be transmitted in much higher carrier frequency ranges than speech or music. The picture is built up of 625 lines in each frame, and frames are produced at the rate of 25 per second. If we assume that we need detail *along* each line on about the same scale as the vertical detail from line to line the total number of changes in information per second must be at least $625 \times 625 \times 25$ which is approximately equal to 10^7. Thus a band width of the order of 20 MHz is needed. The medium wave radio band covers approximately the range 0·3 to 3 MHz and therefore would clearly be useless for conveying T.V. signals as one could hardly imagine modulating a carrier at a *higher* frequency than its own! Nevertheless, for speech with a maximum frequency of, say, 10 000 Hz (bandwidth 0·02 MHz) there is no problem. The U.H.F. band (300–3000 MHz) is perfectly able to accommodate T.V. bandwidths however. You will notice that, in both cases, the bandwidth is about 1 per cent of the carrier frequency in the upper part of the frequency range.

We must now consider how the concept of bandwidth relates to that of temporal coherence. You will recall that earlier in the chapter, when we considered coherence of water waves, I said that the more indefinite the wavelength or frequency, the shorter was the wave train, and we introduced the term ' coherence length ' as a measure of temporal coherence. Now consider a wave which is modulated at a particular frequency. Fig. 2.6 (*c*) shows that the carrier is divided up into groups and that there are a number of carrier cycles in each group. Between each group there is a region of uncertainty in phase and so the length of each group is a measure of the coherence length for the modulated wave. Now the number of carrier cycles in the group *increases* as the modulation frequency *decreases* and so the coherence length is inversely proportional to the modulation frequency, i.e. to the bandwidth.

In round figures, if the bandwidth is x per cent of the carrier frequency then there will be of the order of $1/x$ carrier wavelengths per modulation cycle. In other words the coherence length expressed in carrier wavelengths is approximately the inverse of the bandwidth expressed as a percentage.

How does this work out for T.V. and radio? In both cases we said that the bandwidth is of the order of 1 per cent and so the coherence length is about 100 wavelengths. For T.V. this means about 15 m and, for the medium wave radio, about 15 km. In the visible region the familiar sodium flame has a bandwidth which is of the order of 0·1 per cent and so the coherence length must be of the order of 1000 wavelengths, which is approximately $100 \times 6 \times 10^{-7}$ m $= 6 \times 10^{-4}$ m, i.e. 0·6 mm. A helium–neon laser on the other hand might have a bandwidth of the order of 10^{-8} per cent and this leads to a coherence length of the order of 6 km!

Before leaving this topic it is probably worth pointing out the connection between the ideas of coherence and bandwidth and that great fundamental concept of physics known as the Uncertainty Principle, or Heisenberg's Principle. Crudely put, it states that the more you know of one aspect of a particle the less you know of another. Thus if you know precisely *where* an electron is you can say little about its momentum, whereas if you know its *momentum* precisely you cannot know where it is.

We have been talking about electromagnetic waves and, as you know, the complete mathematical description of their behaviour incorporates some elements of wave-like behaviour and some of particle-like behaviour and neither is a complete description of its own. For the purposes of this book we talk mainly in terms of waves, but the other aspect must not be forgotten. In terms of the discussion of the last few paragraphs, if we had a wave of completely precise frequency—i.e. a plain carrier with no modulation—it would have zero bandwidth and hence infinite coherence length. In photon terms therefore its *location* would be completely indefinite. On the other hand a wave of very *imprecise* frequency, and hence large bandwidth, would have a very short coherence length and could be located much more precisely in space. The momentum of a photon according to the de Broglie relationship is hf/c, where h is Planck's constant, c the velocity of the wave and f the frequency. So the statement of Heisenberg's Principle in relation to electrons can be seen to be equivalent to that for photons, with momentum substituted for frequency.

2.4. *The nature of the scattering process*

We saw in Chapter one, in the experiment with a slide in the projector but no projection lens, that the patch of light produced on the screen must contain, in the form of phase relationships, all the information necessary to image the slide. We also saw that each point on the screen is receiving light—and therefore information—from each point on the slide. Following on from the discussions of the first few sections of the present chapter, we should now be able to understand more clearly how the lack of both temporal and spatial coherence in the projector illumination leads to the lack of any visible interference effects on the screen. The phase relationships are there, however, and the fact that we can form an image with a lens proves the point. However, for the purposes of our present argument, we can consider the source of light in the projector as being made up of an enormous number of independent point sources each of which is small, precisely located, and temporally coherent. The whole source is made up of point sources of all the different wavelengths present in the white light. Each of these separate sources gives a perfectly good interference pattern but each pattern is at a different position on the screen and will be phase-independent of all the others. Hence they add up to give the familiar patch of white light. It will thus be much more profitable for us to consider the pattern produced by a single temporally and spatially coherent source and, if need be, we can consider problems relating to incoherent illumination by adding up the independent effects at a later stage (see p. 54).

Let us therefore consider the geometry of scattering by an object placed in a temporally and spatially coherent beam of light—which, to make it easy to begin with, we will consider to have plane wave fronts. In simpler language it is in a parallel beam of monochromatic light! Fig. 2.7 shows the patches produced by two different objects in spatially coherent monochromatic light. The first object is regular and gives a regular interference pattern which would commonly be called a diffraction pattern. The second object is random and gives a much less regular pattern which might more commonly be described as a scattering pattern. Either term is in fact acceptable at this stage, but we shall discuss nomenclature in more detail in the next section.

We have now reached an awkward point in the development of our subject at which a number of ideas are needed simultaneously and yet clearly they must be developed in sequence. I shall have to ask you to accept one or two statements at face value here and to be patient until the ideas behind them can be developed logically later in the book. The first is simply to note that we have slipped into the habit of talking about diffraction or scattering with visible light. It is so much easier to talk about things that we can see directly and many of our discussions will be conducted using light as the example, but it is important to remember that all the geometrical aspects apply equally well to all kinds of radiation and this may not always be stressed explicitly. The second concerns the detection or recording of radiation; photographic film (or indeed the eye if we are making direct observation) responds to the *intensity* of the radiation (that is to the square of the amplitude) and we can make no record of the relative phases. However, when scattering patterns are superimposed the phase is of importance. All the optical diffraction patterns used in this book are thus records of the distribution of the square of the amplitude of the diffracted radiation. Such patterns are often referred to in the literature as ‘ optical transforms ’. The two-dimensional objects containing apertures used to scatter or diffract light are usually called ‘ masks ’.

The third point concerns the way in which we need to look at the patterns. Figs. 2.7 (*a*) and (*d*) are enlarged prints from masks made up of holes which in the original, are 1 mm in diameter. Figs. 2.7 (*b*) and (*e*) are from strongly exposed negatives, and the central disc surrounded by concentric rings in each is related to the size of hole (the broad features are in fact those of the Airy disc pattern of any one hole, as we shall see later). However, at this point we are only interested in the detail arising from the way in which the holes are distributed; for this reason we can ignore the rings and study only the detail in the central disc. Figs. 2.7 (*c*) and (*f*) are exposed and enlarged to show mainly the detail in this disc and most of the later pictures will be presented in this form.

Now we can return to our main theme. Whatever the object, it is

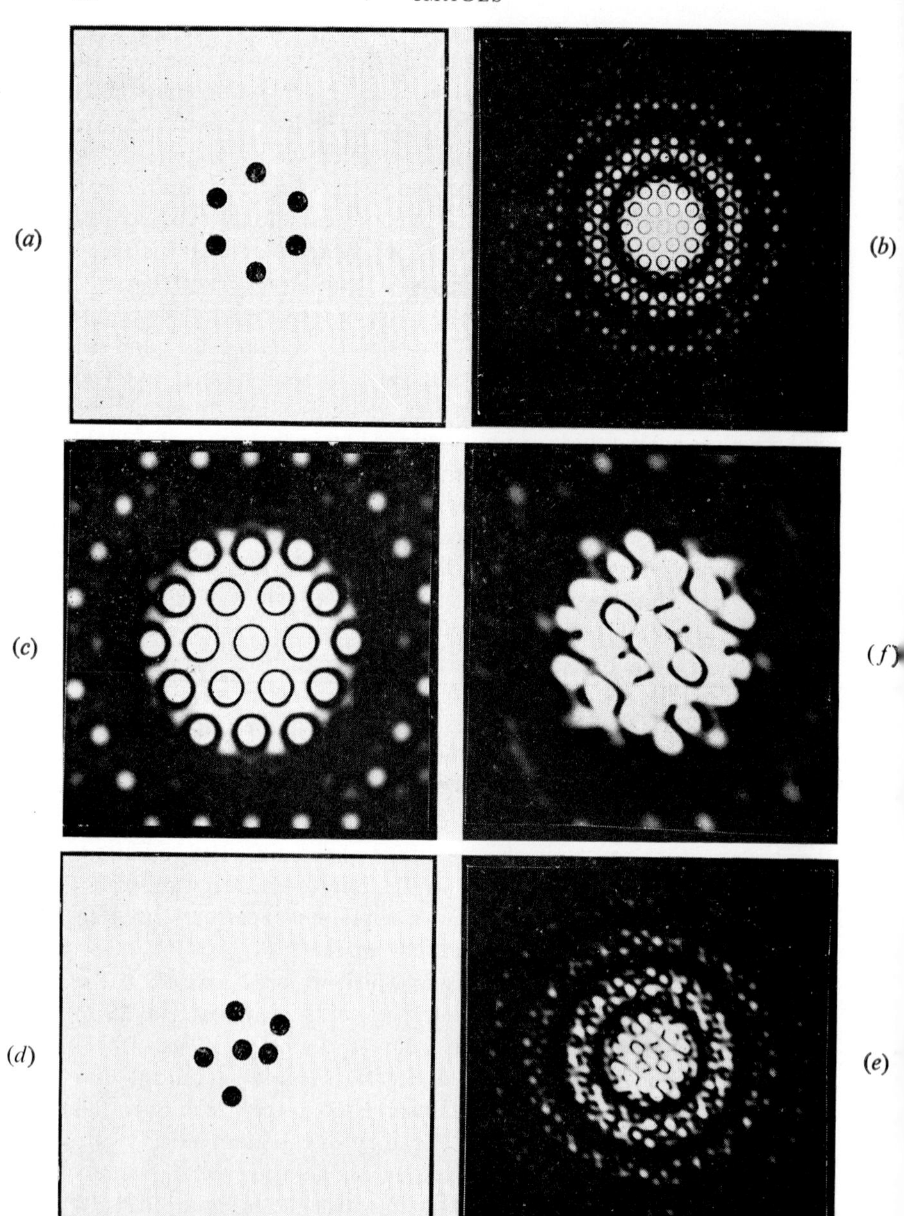

Fig. 2.7. (*a*) Enlarged print from mask of holes each 1 mm in diameter. (*b*) Strongly exposed diffraction pattern of (*a*). (*c*) Diffraction pattern of (*a*) with smaller exposure and greater enlargement than for (*b*), showing the detail within the central disc which is of most interest for our purposes. (*d*), (*e*) and (*f*) as (*a*), (*b*) and (*c*) with a less regular arrangement of holes.

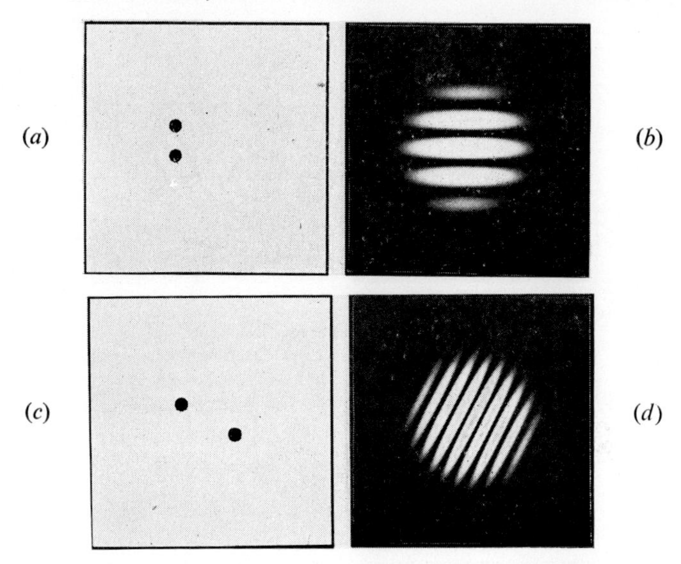

Fig. 2.8. (*a*) Two holes selected from fig. 2.7 (*a*). (*b*) Diffraction pattern of (*a*). (*c*) Another pair of holes selected from fig. 2.7 (*a*) which are twice as far apart as those for (*a*). (*d*) Diffraction pattern of (*c*); the fringes are half as far apart as those in (*b*).

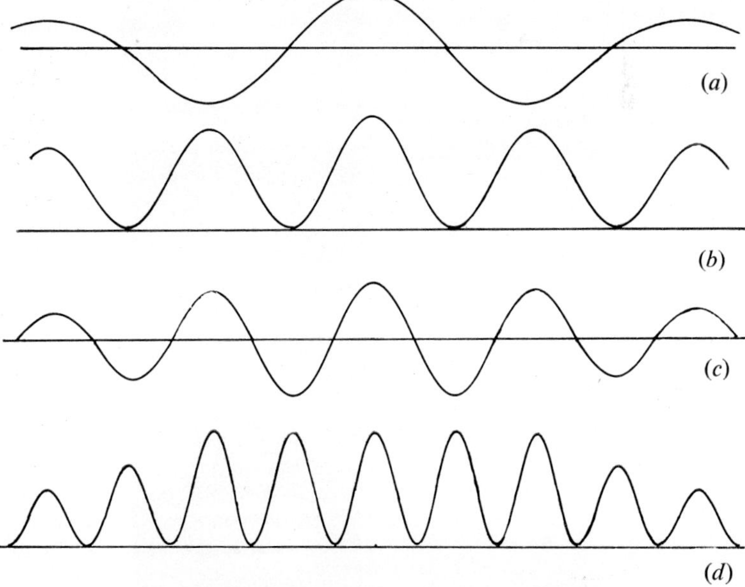

Fig. 2.9. (*a*) Amplitude along a line perpendicular to the fringes of fig. 2.8 (*b*). (*b*) Intensity corresponding to (*a*). (*c*) Amplitude along a line perpendicular to the fringes of fig. 2.8 (*d*). (*d*) Intensity corresponding to (*c*).

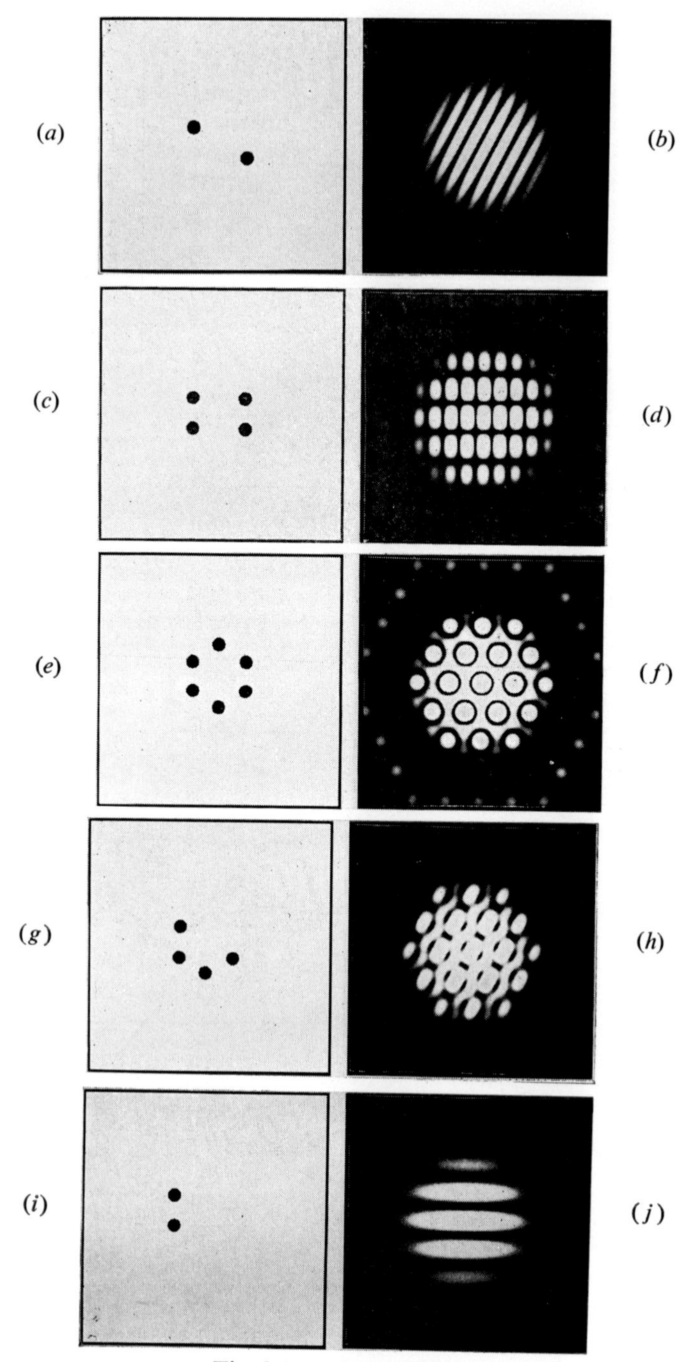

Fig. 2.10 (*cont. opposite*).

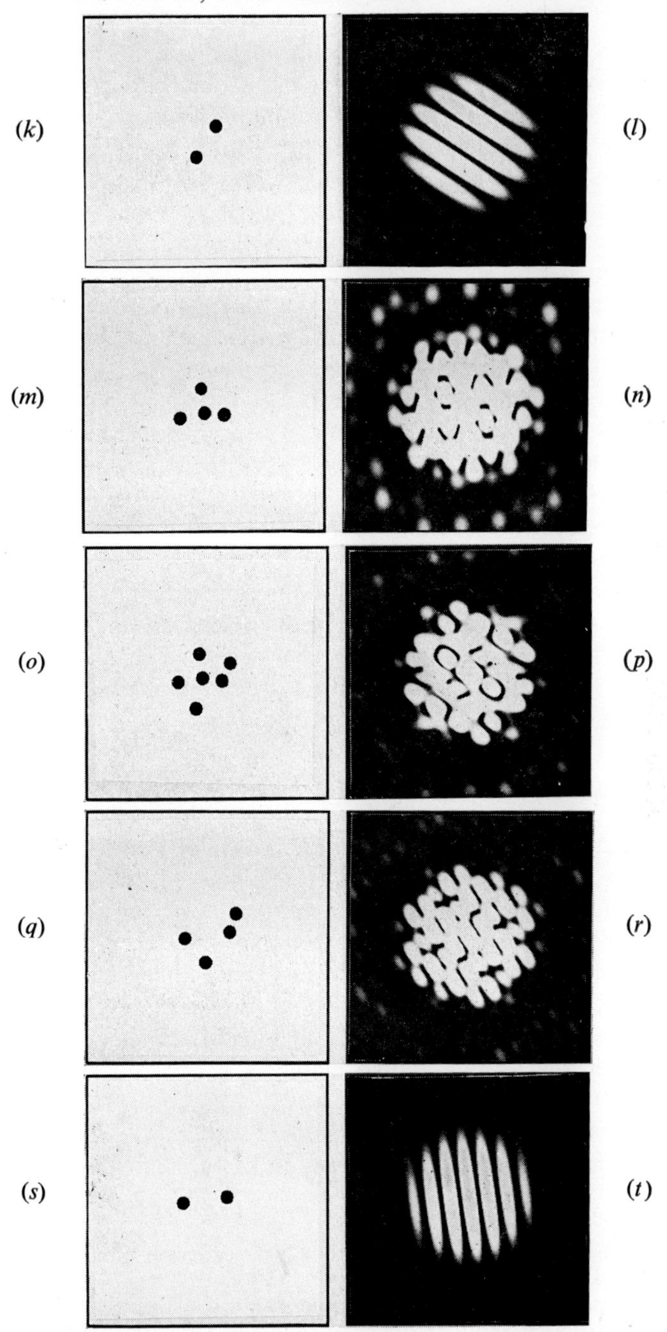

Fig. 2.10. (*a*) and (*i*) Two different pairs of holes. (*c*) and (*g*) Two different selections of 4 holes. (*e*) The complete hexagonal pattern. (*b*), (*d*), (*f*), (*h*) and (*j*) Diffraction pattern of (*a*), (*c*), (*e*), (*g*), (*i*) respectively. (*k*) to (*t*) As for (*a*) to (*j*) for a different arrangement of holes.

possible to consider it to be made up of large numbers of pairs of points. The two objects chosen for fig. 2.7 are particularly easy to divide up in this way but the principle still applies to any object whatever. Every pair of points gives rise to a pattern of fringes which vary sinusoidally in amplitude. Fig. 2.8 (*b*) shows the fringes for the single pair of points at (*a*); Fig. 2.8 (*d*) shows the fringes from a pair of points twice as far apart and in a different orientation, as shown at (*c*). Figs. 2.9 (*a*) and (*b*) show plots of the amplitude and intensity respectively along a line perpendicular to the fringes of 2.8 (*b*); 2.9 (*c*) and (*d*) shows the corresponding plot for 2.8 (*d*).

Fig. 2.10 shows how these fringes add together as successive pairs of holes are added to make the objects used for fig. 2.7. This addition should make clear the point made above that the phase is important in the addition, though what we record is the square of the result. It does not matter which pairs are chosen and we have built up each fringe pattern in two different ways to show that the final result is the same. It is important to remember that these patterns are in spatially and temporally coherent light and so interference occurs between each successive set of fringes and the resultant depends on the relative phases of the parts being added.

Figs. 2.11 and 2.12 show some more complex examples of interference patterns in coherent light resulting from diffraction or scattering by masks which in 2.11 can be readily split up into pairs of holes but in 2.12 are more continuous in form. The mathematical process needed to predict the patterns from objects such as those of 2.7 and 2.11, where discrete pairs can be identified, is that of Fourier *summation*. For 2.12, where more continuous areas are involved, and it is necessary to think in terms of pairs of small elements of area, the process is that of Fourier *integration*. Both of these topics are of enormous importance in diffraction and image theory and, though any kind of mathematical treatment is beyond the scope of this book, some of the physical ideas involved will crop up from time to time later on. A more general term that embraces both mathematical processes is ' Fourier transformation '.

Before leaving the geometrical aspects of scattering we shall consider a slightly more mathematical discussion of the scattering from two points.

* In fig. 2.13, A and B are two equal scattering points, which are supposed to be illuminated by a parallel beam of monochromatic light along the direction OP which is the perpendicular bisector of AB. Thus the two points A and B are illuminated in phase with each other. At the point P on a screen the paths AP and BP are equal in length. Hence

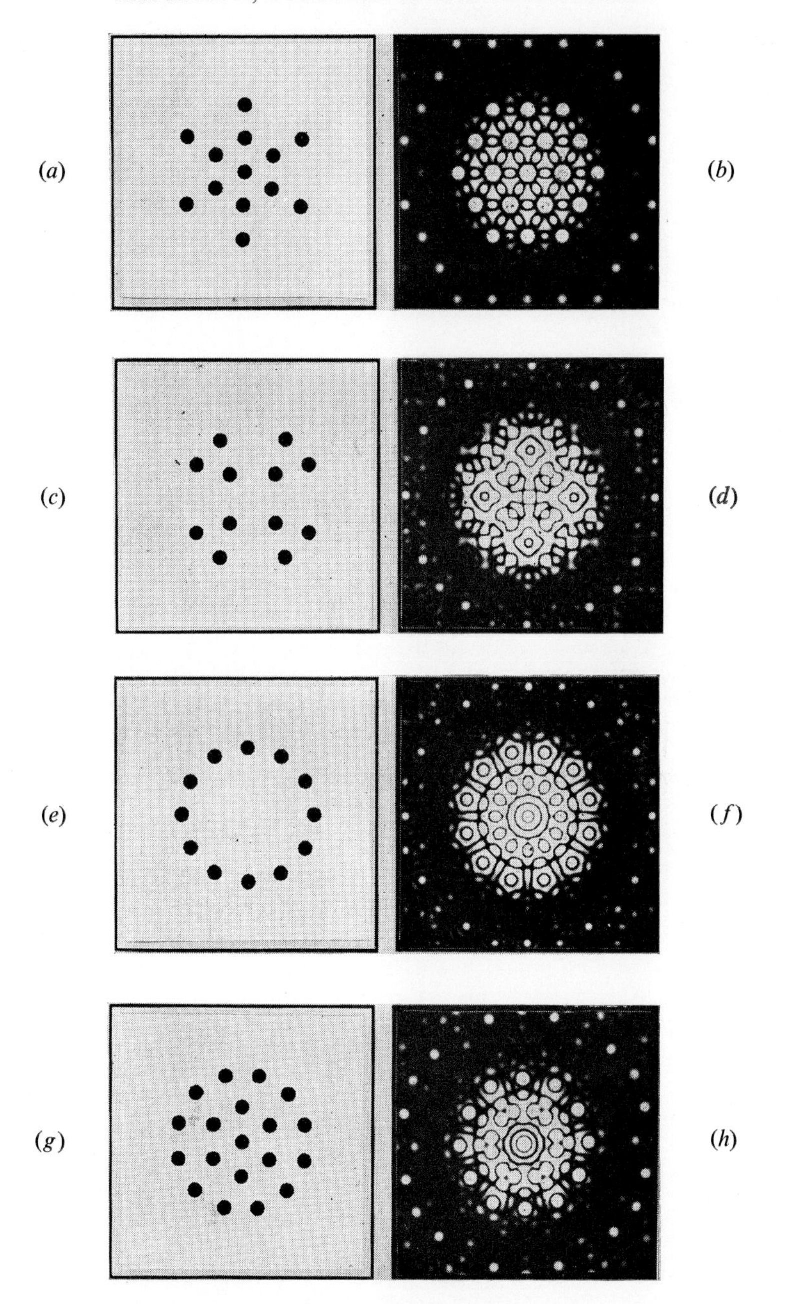

Fig. 2.11. (*a*), (*c*), (*e*), (*g*) Four symmetrical arrangements of 1 mm holes. (*b*), (*d*), (*f*), (*h*) The corresponding diffraction patterns.

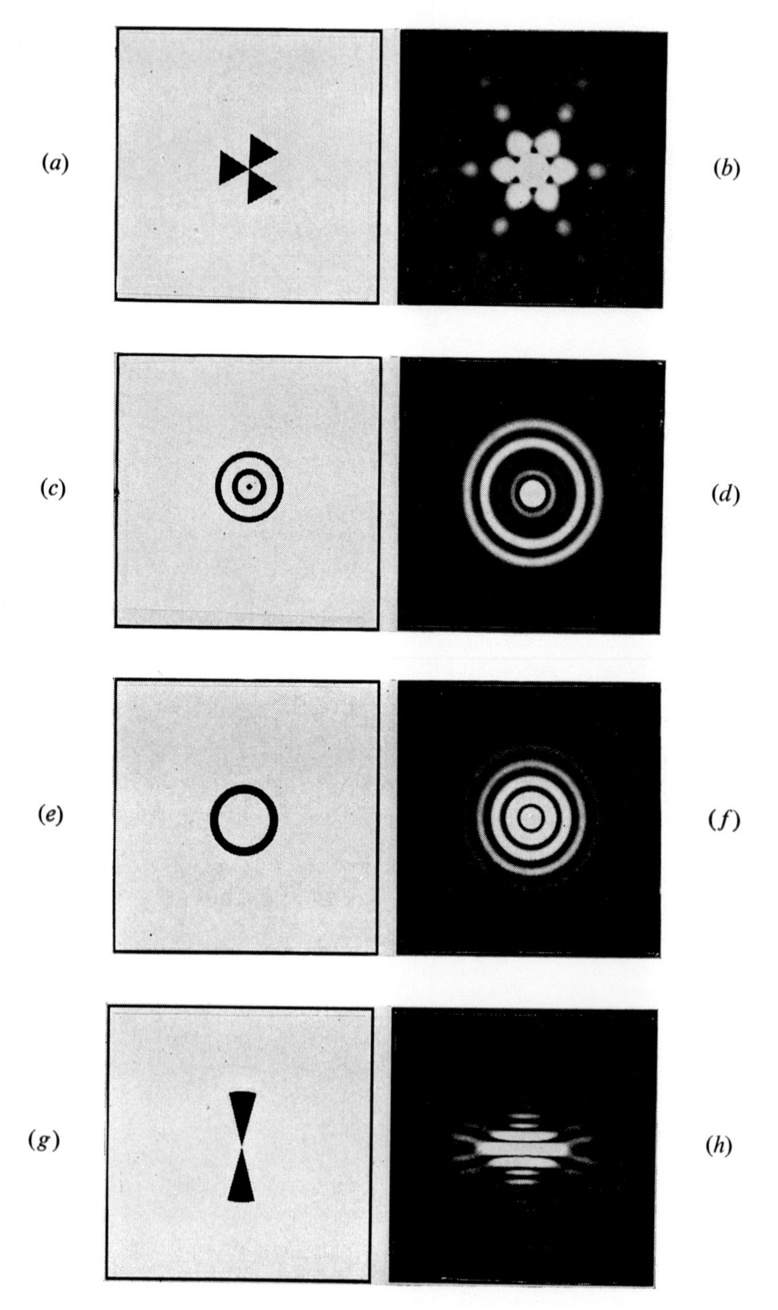

Fig. 2.12. (*a*), (*c*), (*e*), (*g*) Four apertures of different shape. (*b*), (*d*), (*f*) (*h*) The corresponding diffraction patterns.

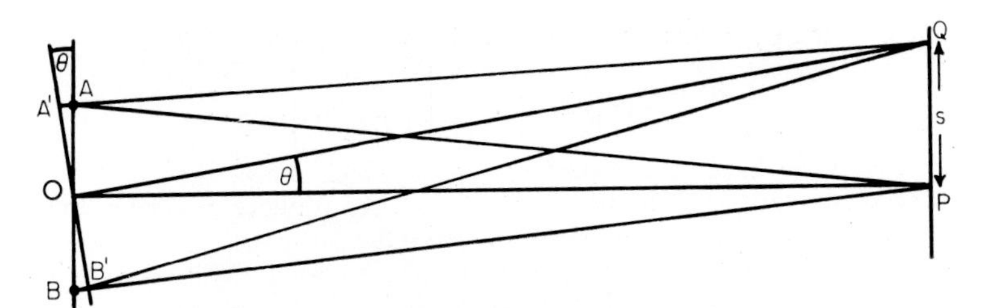

Fig. 2.13. Diagram showing change in path length for scattering at an angle θ to the line joining two sources A and B.

the waves from A and from B arrive in phase and the resultant amplitude is twice that due to A or B alone.

Let us now consider what happens at a point Q on the screen where the angle QOP is θ. Now the wave from A has a shorter path AQ than the wave from B, which is BQ. If we construct an isosceles triangle A′QB′ with its base passing through O it becomes clear that the effective path difference for the waves from A and B to Q is AA′ plus BB′. If θ is small, and if OP is very large compared with AB—both of which conditions are usually met in practice because the screen is placed a long way from the slits—then $AA' = BB' = \frac{AB}{2} \sin\theta$ and the total path difference is $AB \sin\theta$. Thus the phase lag of the wave from B relative to that from A, which we will call 2δ, is $\frac{2\pi}{\lambda} \times$ the path difference, i.e. $2\delta = \frac{2\pi}{\lambda} AB \sin\theta$.

Now let the distance $QP = s$ and the distance $OP = D$. Then, again if D is very large and θ is very small, $s = D \sin\theta$.

Let AB, the separation of the scattering points, be d. Then $\delta = \frac{\pi}{\lambda} AB \sin\theta = \left(\frac{\pi}{D\lambda}\right) ds$, and of course $\frac{\pi}{D\lambda}$ is a constant for a given experiment.

We now need to find the resultant amplitude at Q from two disturbances of equal amplitude but with a phase difference 2δ between them. Without going into detail of the form of the disturbance in electromagnetic terms, we can assume that it will be a sinusoidal wave of some kind. For convenience, and in order to retain symmetry—which turns out to be useful later—let us assume that *if* there had been a scatterer at O it would have created a disturbance at Q of the form

$$\phi_0 = a \sin(2\pi ft),$$

where ϕ_0 is a measure of the disturbance at Q at a time t, a is the amplitude and f the frequency. Then the disturbances from A and B will be

$$\phi_A = a \sin(2\pi ft + \delta) \text{ and}$$
$$\phi_B = a \sin(2\pi ft - \delta).$$

So the sum which we require is

$$\begin{aligned}\phi_A + \phi_B &= a \sin(2\pi ft) \cos\delta + a \cos(2\pi ft) \sin\delta \\ &\quad + a \sin(2\pi ft) \cos\delta - a \cos(2\pi ft) \sin\delta, \\ &= 2a \sin(2\pi ft) \cos\delta = 2\phi_0 \cos\delta.\end{aligned}$$

From our earlier work this can be written

$$\phi_A + \phi_B = 2\phi_0 \cos\left[\left(\frac{\pi}{D\lambda}\right) ds\right].$$

Thus we have two important results which are at the heart of all diffraction theory. The first is that, since the product ds occurs on the right-hand side, it follows that if we *reduce* the slit spacing (d), the spacing of the points of similar amplitude on the screen (s) will *increase* and vice versa. The second is that the variation in amplitude is of cosine form. Thus since any real object can be considered to be made up of many pairs of scattering points, any diffraction pattern can be considered to be made up of many cosinusoidal fringes.

Before leaving this discussion it may be useful for future purposes to note that the resultant can be found using a geometrical construction called the 'phasor diagram'. A phasor has exactly the same combinatorial properties as a vector but is a purely constructional device which has no physical entity as does a real vector.

The diagram we should use is as shown in fig. 2.14. The phasors representing the amplitudes transmitted by A and B, which are respectively a phase angle δ ahead of a hypothetical wave from O, and δ behind, are shown as thick arrows. The resultant obtained by completing the 'parallelogram of vectors' is the dotted arrow. In each case it is easy to verify that this is a geometrical solution of the mathematical result $\phi_A + \phi_B = 2\phi_0 \cos \delta$ since the *magnitudes* of ϕ_A, and ϕ_B are both equal to the magnitude of ϕ_0.

The phasor-diagram technique is valuable in solving many diffraction and interference problems. For a complete mathematical discussion of the concept a text-book on optics should be consulted (e.g. *Optics* by Smith and Thomson).

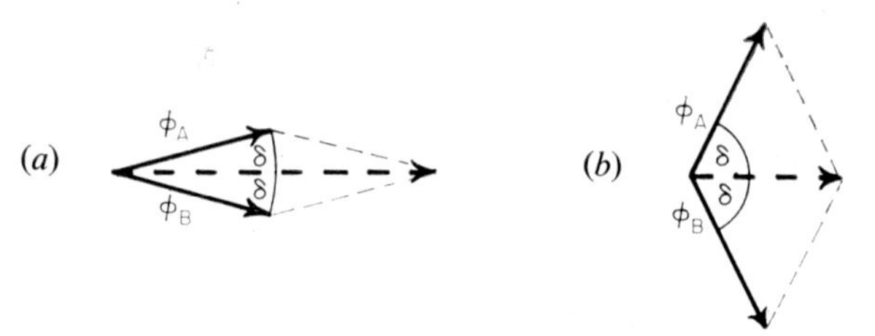

Fig. 2.14. (*a*) and (*b*) Phasor diagrams illustrating the resultant scattering as the phase angle δ changes.

It is perhaps important to remind ourselves that all these geometrical considerations apply whatever the radiation; the theoretical formulation—apart from problems of scale—is the same whether we are considering visible light, radar or acoustic waves. There are however physical factors that affect the phase relationships and the most obvious is one that leads to a change in velocity of the wave. In optics we think of this in terms of the refractive index of the medium; the refractive index is simply the ratio of the velocity of the waves in free space to that in the medium. (The concept applies equally to other forms of radiation and we may quite properly talk of the refractive index for ultrasonic waves or for radio waves.) Suppose in a simple double slit experiment, one of the slits is covered with a thin plate of a medium of a refractive index greater than 1. If its thickness is t and the refractive index n, the additional optical path length of the waves going through this slit relative to the other is $(n-1)t$ and so the net result is to move the centre of the pattern over to one side to a point at which the phase difference due to the geometrical effect discussed in the last section compensates for the phase difference due to the plate. A little thought shows that the whole pattern moves *en bloc.* (But see also the end of Section 2.8 for a complication that occurs if the light is not highly monochromatic.)

This is a relatively easy problem to deal with, but if the object causing the additional path is irregular then the situation can be very difficult. When one considers that the wavelength of light is of the order of 5×10^{-7} m it is obvious that a change in thickness of a piece of glass of refractive index $\sim 1{\cdot}5$ of the order of only 0·5 microns would lead to a phase difference of 180°. In producing diffraction patterns, therefore, it is usual to use slits and holes rather than black and white patterns on a photographic plate. The masks used for Figs. 2.11 and 2.12 were produced by making holes in thin, opaque, ‘ estar ’ sheet and the contact prints are approximately actual size.

2.5. *Nomenclature of diffraction and scattering processes*

The nomenclature of scattering and diffraction processes is highly confused, and in order to make progress the only real possibility is to define a set of terms that are self consistent—which is more than some text books do, I regret to say—and then to stick to them even though they may differ from those used by other authorities. I prefer to keep the term diffraction to describe any passive interaction between radiation and an object; thus in my terminology scattering includes diffraction but may also include active interactions (such as for example Compton scattering where a change of wavelength occurs). Since discussions of these forms of scattering are not included in this book, it follows that for our purposes scattering and diffraction are synonymous.

Interference is the interaction of radiation with itself, i.e. without the presence of material objects. Thus, in my terminology, in the familiar Young's double-slit interference experiment, scattering or diffraction occurs at the first single slit and again at the pairs of slits: the resultant wave trains then *interfere* to produce the pattern. In a Newton's rings experiment the various trains of waves are created by multiple reflection and transmission and their subsequent interaction is interference. In a diffraction grating the incident radiation is scattered or diffracted by the grating and the subsequent wave trains interfere to produce the pattern. If this scheme is followed it leads to unambiguous descriptions of the various processes.

My only real regret is the need to use the term ' interference ' for the interaction of waves. In linear systems the principle of superposition is obeyed—that is the resultant disturbance at any point is the sum of all the separate disturbances at that point—and so the waves have no permanent effect on each other and so ' non-interference ' would be a better term. However, the term has been in use too long to change now! What happens, of course, is that a collection of waves passing through a point lead to a certain resultant at that point *as viewed by an outside observer*, but we must remember that even at a null point in an interference pattern, waves are continuously passing through and conveying energy; it is merely the external combined result that is zero. The effect is somewhat like that of a man climbing up an escalator at the same speed as it is descending; to an observer some distance away who cannot see his legs he appears stationary—though clearly this is the result of two movements which neutralize each other as far as the outside observer is concerned.

In classical treatises on diffraction it is customary to divide phenomena into two categories which are usually described as ' Fraunhofer ' and ' Fresnel '; the impression is sometimes created that there are just two patterns that any particular object can produce—one in the Fraunhofer category and the other in the Fresnel category. There is however an infinite number of patterns and the two often described in the literature are limiting cases.

Fraunhofer diffraction is quite specific and we can speak of ' the ' Fraunhofer diffraction pattern of an object to describe the pattern resulting from subsequent interaction of the waves when radiation is scattered by the object under the following conditions:

(1) The radiation is monochromatic, i.e. is temporally coherent;
(2) The incident radiation has plane wave fronts and is laterally coherent;
(3) The subsequent interference pattern is viewed at infinity—or in the back focal plane of a converging lens receiving the scattered radiation, which of course is equivalent to viewing at infinity.

Under these conditions the *only* readily observable variations in pattern (not counting absolute intensity level—which depends on the intensity of the incident radiation) for a given object are of scale and of extent and these factors depend on the relative scale of the object and wavelength. The cautious phrase ' readily observable ' is inserted because, although the statement is absolutely true for the *intensity* distribution in the pattern (which depends on the square of the amplitude), the phase distribution depends on other factors which we have not included here. However, the phase can be observed only in the special circumstances of the addition of a coherent beam as in holography (see Chapter 6) so the effect need not worry us unduly.

Reference was made earlier to the fact that the mathematical relationship between an object and its scattering or diffraction pattern involves Fourier transformation. In more precise terms, if we add to the three conditions for Fraunhofer diffraction specified above, a fourth:

(4) The diffracting object is placed in the *front* focal plane of the converging lens specified in (3),

then the distribution in its back focal plane (assuming that the lenses are fully corrected for aberrations) is exactly described by the Fourier transform of the object in terms of both amplitude and phase. I suppose, strictly speaking, that this is the only true Fraunhofer diffraction pattern.

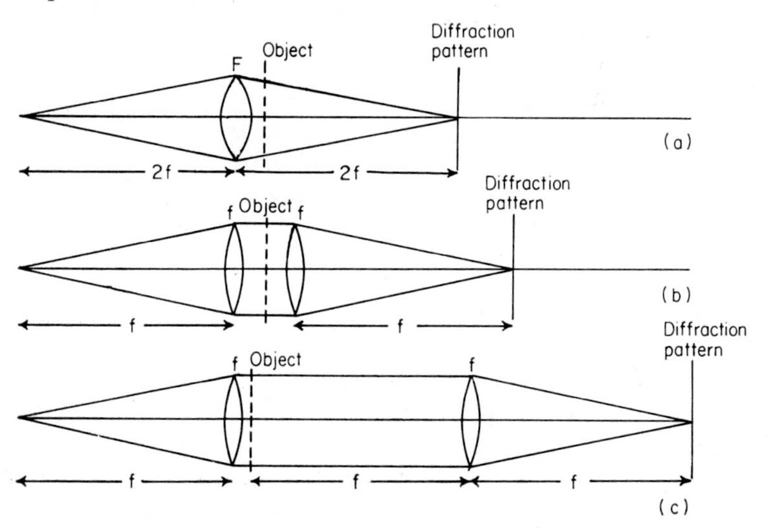

Fig. 2.15. (*a*) Simple arrangement for observing Fraunhofer diffraction which approximates to conditions 1, 2 and 3 on page 36. (*b*) Arrangement for observing Fraunhofer diffraction that conforms with conditions 1, 2 and 3 on page 36. (*c*) Arrangement for observing Fraunhofer diffraction that also satisfies condition 4 of page 37 and which therefore gives true Fourier transforms.

However, since we are normally concerned only with intensities, particularly if we are taking photographs, this restriction need not be applied rigorously. Indeed, the patterns produced under conditions 1, 2 and 3 above have already been described as optical transforms to distinguish them from precise Fourier transforms. The simple set-up often used in school laboratories, shown schematically in fig. 2.15 (*a*) is a reasonable approximation to the arrangement shown in fig. 2.15 (*b*) which *does* conform with conditions 1, 2 and 3. It will certainly give Fraunhofer diffraction patterns, or optical transforms, whose *intensity* distribution will be virtually indistinguishable from those produced by the arrangement of fig. 2.15 (*b*), and also from the patterns produced by the arrangement of fig. 2.15 (*c*) which is the arrangement satisfying all four conditions, and hence giving correct phases.

It is convenient to describe all the patterns produced when conditions 2 and 3 are not obeyed as Fresnel patterns—though there are of course an infinite number associated with each object depending on the nature of the incident wave fronts and of the viewing system. As an example we give in figs. 2.16 (*b*)–(*f*) the Fraunhofer diffraction pattern and four different Fresnel patterns of the same hexagonal aperture.

In most of the applications described in this book the object will be far enough away from the source of radiation for the wave fronts to be nearly parallel and the viewing arrangements will not depart much from condition (3) above and so the intensity distribution in the relevant pattern will be little different from the Fraunhofer pattern. Mathematically speaking we often describe this as the ‘far-field approximation’. There is a mathematical formulation which applies in general but which is greatly simplified when the three conditions are fulfilled and the name derives from this.

Finally it is important to notice that there is often confusion over the differences between so-called one, two and three dimensions in relation to diffraction. In fact, of course, diffraction can only be thought of in three dimensions. The most useful concept in interpreting diffraction is that of Huygens' principle and this depends on the idea that each part of the wave front acts as the centre of a new *spherical* wave; this immediately means that one must think in three dimensions. We shall see later that, whereas X-ray diffraction is often claimed to be essentially three-dimensional because one normally thinks of the interaction with the atoms in a crystalline solid, optical diffraction is usually thought of in terms of the interaction with a planar object. We shall see that the significant differences between X-ray and optical diffraction do not arise from this point but from the difference in the relative scale of the radiation and of the object: X-rays have wavelengths comparable with

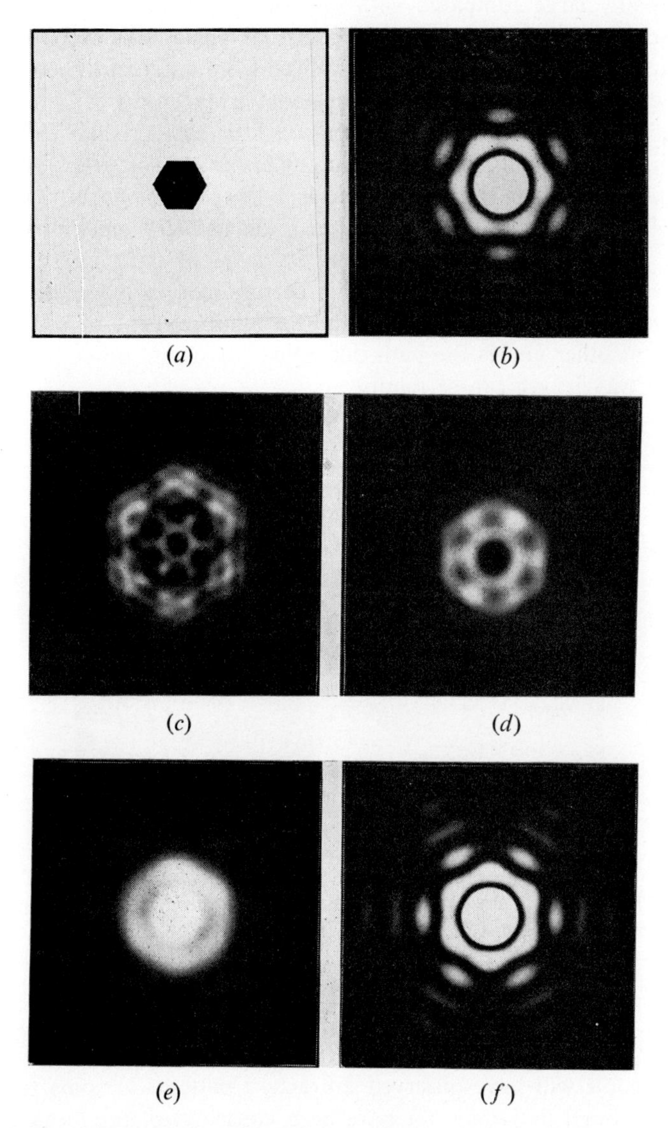

Fig. 2.16. (*a*) Contact print of hexagonal aperture. (*b*) Fraunhofer diffraction pattern of (*a*). (*c*), (*d*), (*e*) and (*f*) Four different Fresnel diffraction patterns of (*a*). The scales of the five patterns have been adjusted to make them all comparable in size.

the atomic separation; objects used in optical diffraction experiments are usually large compared with the wavelength of light. One particularly beautiful exception is the gemstone opal. The ever-changing colours produced when an opal is viewed from different directions are diffracted beams from a three dimensional arrangement of transparent equal silica spheres whose diameter is a little smaller than the wavelength of light. Fig. 2.17 is a scanning electron micrograph showing the spheres. Their refractive index is 1·45 so their *optical* diameters are $1{\cdot}45 \times 2{\cdot}4 \times 10^{-7}$ m $= 3{\cdot}48 \times 10^{-7}$ m. The path difference introduced when light is scattered back along its incident path by successive spheres is thus $6{\cdot}96 \times 10^{-7}$ m. This corresponds to a wavelength in the red region of the spectrum and so reinforcement of red would occur; at other angles the path difference is less and other colours of shorter wavelength can be reinforced.

Fig. 2.17. Scanning electron micrograph of a fragment of opal at a magnification of 5000. The spheres are of transparent amorphous silica about 240 nm in diameter. Photo by Mrs. C. Winters, Department of Zoology, University College, Cardiff.

2.6. *How can we observe optical diffraction patterns?*

Most people will have observed diffraction patterns at some time or other—though they may not have been conscious of the fact. Look out of the window of a bus at a street light when the window is misted or spotted with rain and you will see halos round the light; clean the window by rubbing your glove *horizontally* across the glass and you see *vertical* streaks spreading out from the light; look through an umbrella at a distant street light and you will see a pattern of spots arranged in a regular way. These are all diffraction patterns and indeed are very close to being *Fraunhofer* diffraction patterns as far as their intensity distribution goes. The reason is that the light source in each case is fairly distant and hence the incident light is almost parallel; if the eye is focused on the *source*, then the image on the retina is in the back focal plane of the lens of the eye and so the Fraunhofer conditions are very nearly obeyed.

In my terminology, the familiar Young's double-slit experiment is really a diffraction experiment; the slits interact with the wave fronts and produce diffracted waves which subsequently interfere to produce the fringe pattern. In the laboratory it has been the practice for many years to use slits because this enables us to produce brighter patterns. Slit sources do however obscure the essentially three-dimensional nature of the diffraction process which was mentioned towards the end of the last section. Most of the illustrations in this volume will therefore be produced with *point* rather than *slit* sources but, in order to underline the relationships with experiments that may be more familiar, fig. 2.18 shows a comparison of the patterns produced by two different double-slit spacings and two different slit widths, using both a slit source and a point source.

The simple experiment with a street lamp forms the basis of the simplest available method of observing diffraction patterns in the laboratory. Set up a simple pea-lamp or flash-lamp bulb at one end of the laboratory and view it, with the eye focused on it, through a fine handkerchief, a piece of umbrella fabric or a silk scarf. The regular pattern of spots is related to the fine structure of the fabric. The pattern will, of course, be coloured if white light is used, as there will be a pattern on a different scale for every wavelength present; the overlap clearly produces a coloured result. A piece of green cellophane over the lamp produces sufficiently ‘ monochromatic ’ light to permit the individual pattern for one wavelength to be isolated fairly easily. Replace the regular fabric by a piece of knitted fabric or of a pair of ‘ stretch nylon ’ tights and the regularity disappears and a somewhat diffuse halo is seen. This is still a diffraction pattern and is related in the same mathematical way to the sub-detail of the fabric. Figs. 2.19 (*b*) and (*h*) show photographs of the diffraction patterns of two such pieces of material.

Stretch the material in the horizontal direction and notice that the horizontal dimensions of the pattern *shrink* (figs. 2.19 (*d*) and (*j*)). Then tilt the unstretched fabric so that the light passes through it at about 60° to the normal. Now the projection of the fabric along the light direction has shrunk to half its original dimensions ($\cos 60° = \frac{1}{2}$)

and the details of the diffraction pattern expand to *twice* the size (figs. 2.19 (*f*) and (*l*)).

As a final experiment with this simple arrangement translate one of the pieces of fabric from side to side as you observe its diffraction pattern. The pattern does *not* move, although if there are variations in the regularity or spacing of the threads the pattern may change slightly. These experiments illustrate some of the very important properties of diffraction patterns and (see section 2.3) hence of some of the corresponding mathematical formulations—Fourier transforms.

It may be worth considering for a moment the explanation of these phenomena. The reciprocal effect—that the pattern shrinks or expands when the object expands or shrinks—can be understood most easily by referring back to section 1.2 and fig. 1.3. The striped string experiment shows that when points in the object are *close together* one has to move *further away* from the centre line to find the same phase difference and vice versa.

The lack of movement of the pattern when the object is moved is not difficult to understand. The image of the lamp is focused at one point of the retina and unless the *lamp* is moved relative to the eye this point will always be at the centre of the pattern. All waves diffracted in a particular *direction* in space that are parallel with each other will come to a focus on the retina at a fixed point some distance from the centre and, since lateral translation of the object moves all these waves parallel to themselves the actual pattern does not move.

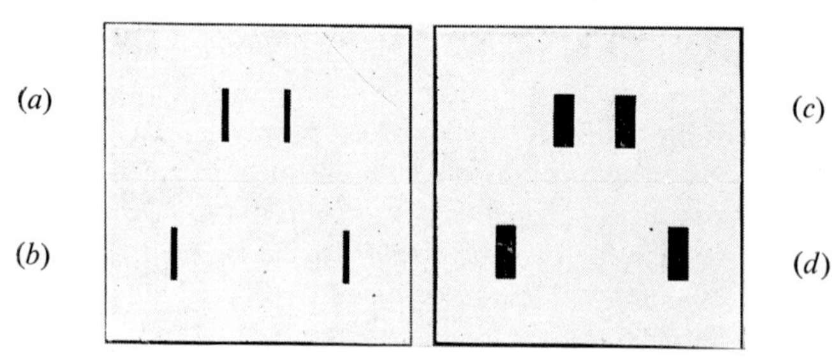

Fig. 2.18 (*cont. opposite*)

Fig. 2.18. (*a*) Two rectangular apertures. (*b*) Two rectangular apertures further apart. (*c*) Two rectangular apertures with spacings as for (*a*) but each wider. (*d*) Two rectangular apertures with spacings as for (*b*) but each wider. *Opposite:* (*e*) Diffraction pattern of (*a*) with slit source. (*f*) Diffraction pattern of (*a*) with point source. (*g*) Diffraction pattern of (*b*) with slit source. (*h*) Diffraction pattern of (*b*) with point source. (*i*) Diffraction pattern of (*c*) with slit source. (*j*) Diffraction pattern of (*c*) with point source. (*k*) Diffraction pattern of (*d*) with slit source. (*l*) Diffraction pattern of (*d*) with point source.

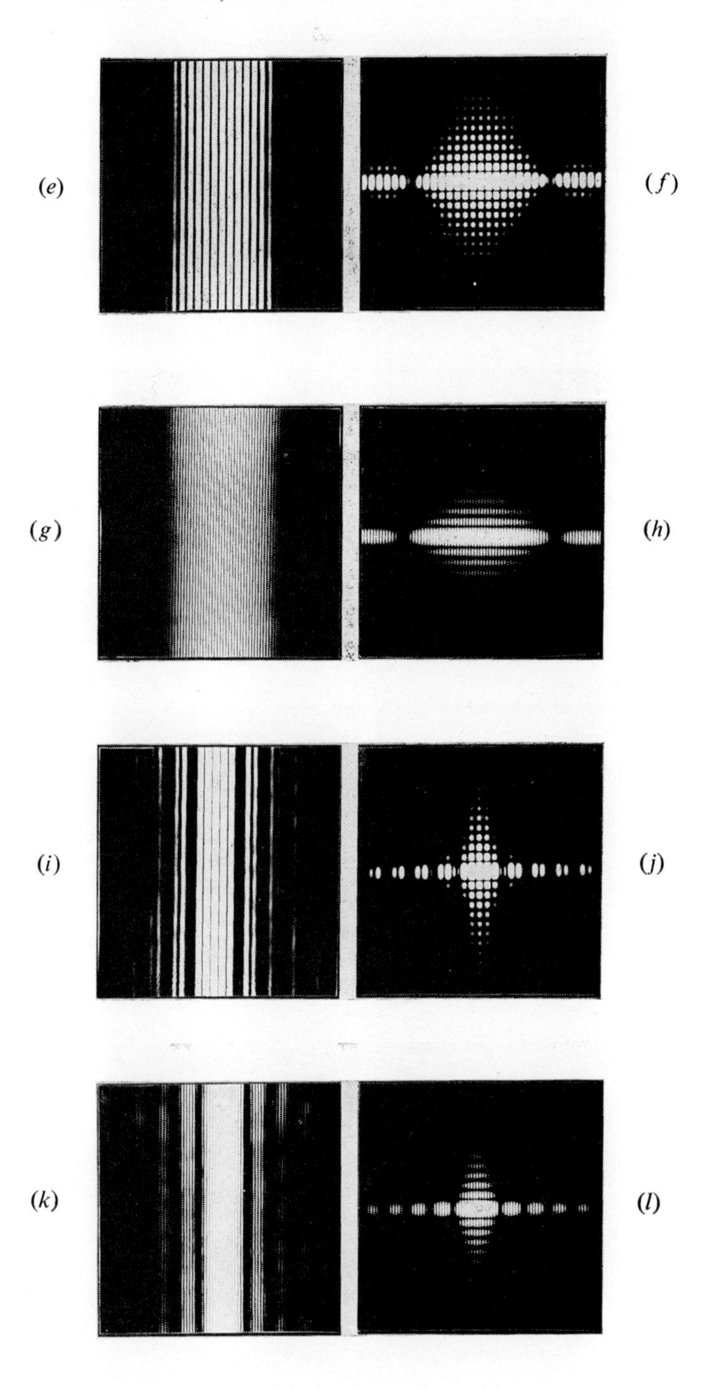
(e)
(f)
(g)
(h)
(i)
(j)
(k)
(l)

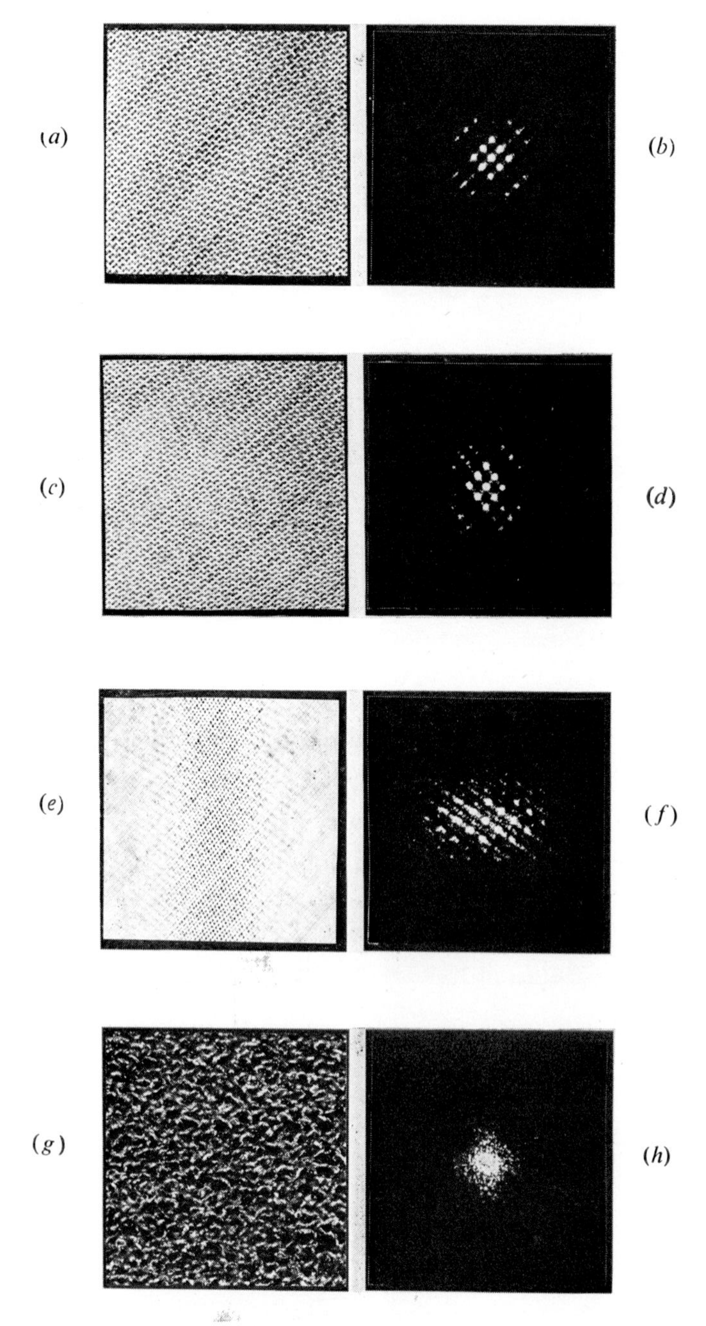

Fig. 2.19 (*see page* 46).

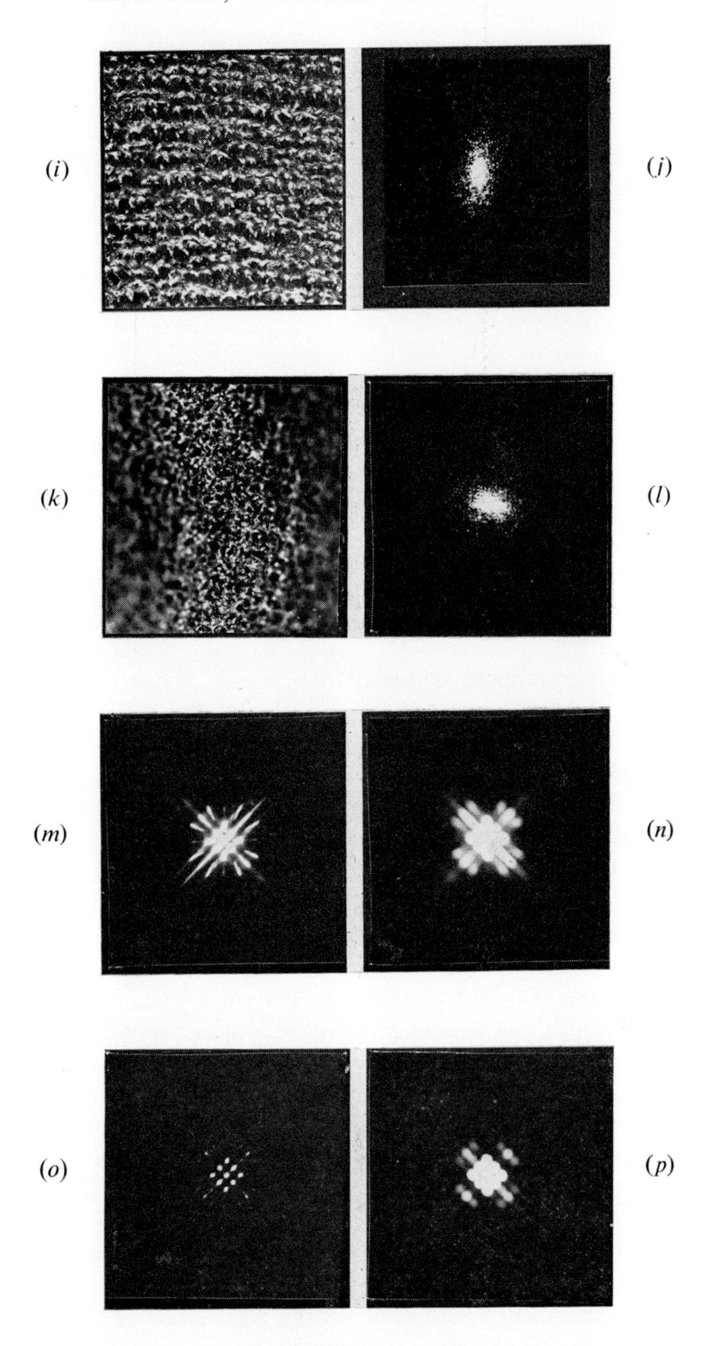

Fig 2.19. (*see page* 46)

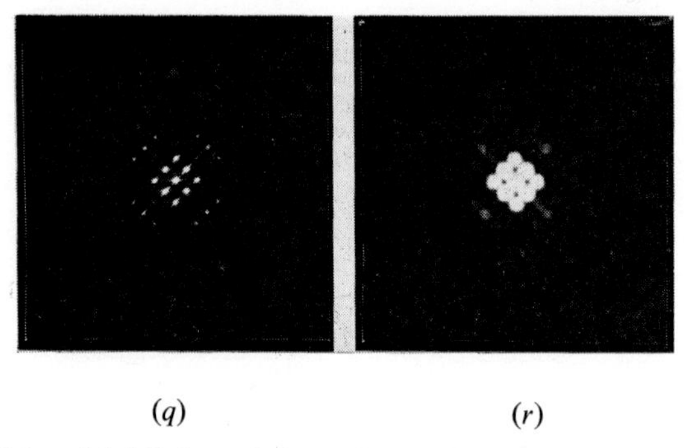

(*q*) (*r*)

Fig. 2.19. (*a*) and (*g*) Enlarged view of umbrella fabric and of stretch tights. (*b*) and (*h*) Diffraction patterns of (*a*) and (*g*) in laser light. (*c*) and (*i*) The fabrics of (*a*) and (*g*) stretched horizontally. (*d*) and (*j*) Diffraction patterns of (*c*) and (*i*) in laser light. (*e*) and (*k*) The fabrics of (*a*) and (*g*) tilted out of the plane of the page. (*f*) and (*l*) The diffraction pattern of (*e*) and (*k*) in laser light. (*m*) to (*r*) are all diffraction patterns of fabric (*a*): for (*m*) and (*n*) the source of light is white and is small and large respectively; for (*o*) and (*p*) the source is green and is small and large respectively and for (*q*) and (*r*) the source is a He–Ne laser and the source is small and large respectively.

The effects of reduced spatial coherence can, of course, be demonstrated by enlarging the source and those of increased temporal coherence by changing from white to green light. A greatly improved way of observing the same phenomena is to place the fabrics in the unmodified beam of a helium–neon laser. The patterns will then be seen on the screen on which the laser beam ultimately falls. Fig. 2.19 illustrates these various effects using the umbrella fabric of fig. 2.19 (*a*) as the diffracting object.

For research purposes a somewhat more sophisticated system is necessary and fig. 2.20 (*a*) shows a diagram of the system used in the author's laboratories at Cardiff. Fig. 2.20 (*b*) is a photograph of the actual apparatus. A carefully designed optical system L_E (which involves three stages) expands the laser beam, without losing its spatial coherence, so that a uniform patch of light falls on lens L_1 after which the light becomes parallel. Between L_1 and L_2 we thus have a very accurately uniform set of plane parallel wave fronts which will fall on any object placed at P. Lens L_2 (in the absence of an object at P) produces a minute focal spot of intense brilliance in its back focal plane at Q. If an object is placed at P, then this spot (strictly, an Airy disc) becomes the Fraunhofer diffraction pattern of the object.

For the experiment described in sections 5.1, 5.2 and 6.6 an additional lens L_3 may be placed in position in such a way as to produce an *image*

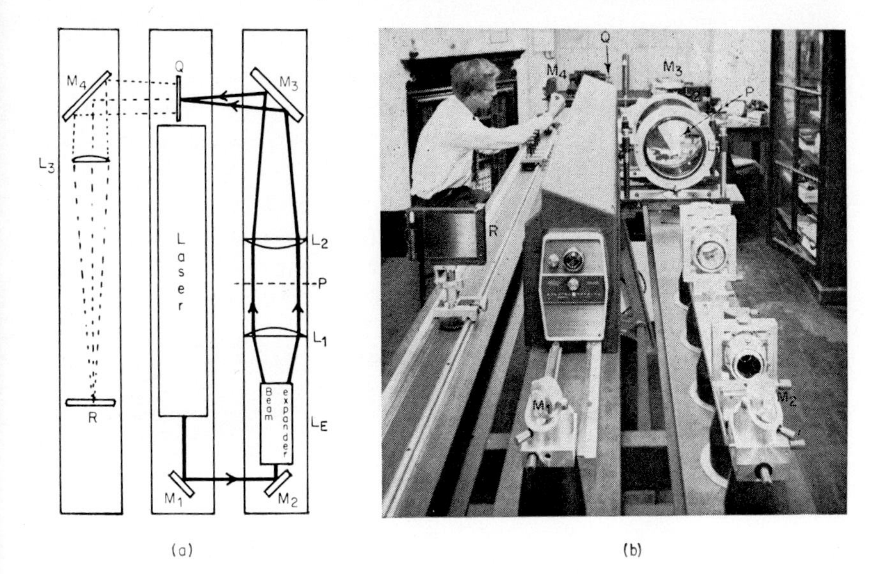

Fig. 2.20. (*a*) Diagram of diffraction system in the Department of Physics at University College, Cardiff (the diffractometer). (*b*) Photograph of the system outlined in (*a*).

of the object P on a screen R. We now have a complete demonstration of the two stages of image formation; from P to Q we have scattering and from Q to R we have recombination. With a 50 mW helium–neon laser this system is capable of producing, for example, a diffraction pattern of a 1 cm diameter hole which has its central disc a centimetre or so in diameter on the screen but is quite bright enough to be seen with the naked eye in a darkened room.

Simple objects in the diffractometer give useful demonstrations of Fraunhofer diffraction principles and with complex objects extremely beautiful patterns can be produced. A selection illustrating different points is shown in figs. 2.21 and 2.22.

This system was developed primarily for use in interpreting X-ray diffraction patterns and the applications are discussed in rather more detail in sections 4.2 and 6.10.

2.7. *How can we record scattering patterns for non-visible radiations?*

So far we have tended to use optics only for our illustrations. The reason is simply that optical patterns are by far the easiest to record. A suitable photographic film enables one to record instantly the whole two-dimensional distribution of intensity, which depends on the square of the amplitude. But we should now examine more closely the fact that it is only a quantity that depends on the *amplitude* that we can record; the phase is not recorded. Indeed our eyes are not sensitive to phase, nor is any other system of recording in the ordinary way. (The exception is in laser holography—a special case which we discuss in Chapter 6.)

Why can we not record the phase differences? Consider yellow light with a wavelength of $5{\cdot}9 \times 10^{-7}$ m (sodium light). The corresponding frequency, if we take the velocity of light to be 3×10^{8} m s^{-1}, is about 5×10^{14} Hz. In other words a phase difference of half a cycle

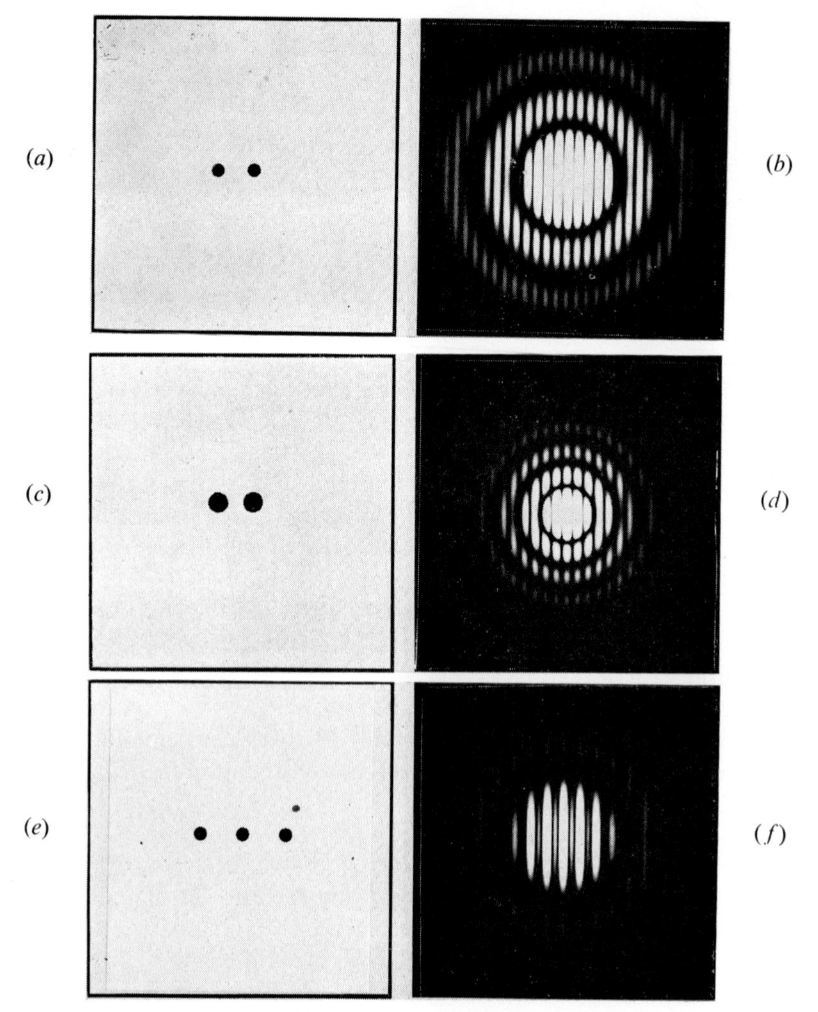

Fig. 2.21 (*cont. opposite*).

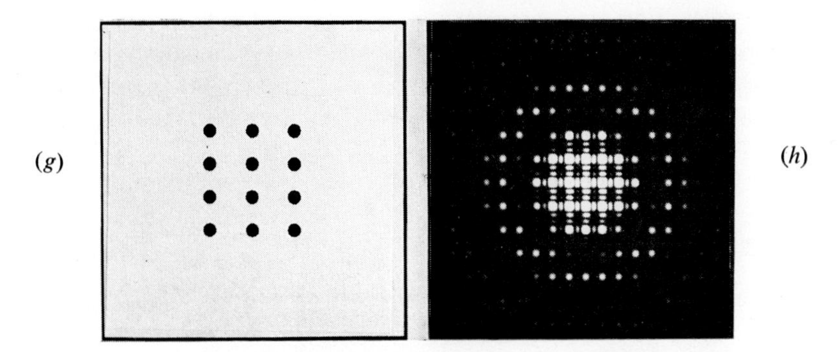

Fig. 2.21. (*a*), (*b*) Two holes and their diffraction pattern exposed to show rings. (*c*), (*d*) Two larger holes with the same spacing as (*a*) and their diffraction pattern exposed to show rings. (*e*), (*f*) Three holes and their diffraction patterns. Note subsidiary maximum. (*g*), (*h*) Twelve holes and their diffraction patterns. Note subsidiary maxima.

Fig. 2.22. (*a*) and (*c*) Masks chosen to give aesthetically attractive diffraction patterns (*b*) and (*d*). From *An Atlas of Optical Transforms*, by G. Harburn, C. A. Taylor and T. R. Welberry, by permission of G. Bell & Sons, Ltd.

involves a time measurement of 10^{-15} second. This is certainly smaller than any time measurement of which our technology is capable at the moment. The same is not true of course for other radiations. In the case of radio waves, phase differences can be measured as indeed they can for sound waves—but of course photographic films cannot be used for them. Thus in order to record patterns of radio or sound waves we usually move a detector from point to point in the area concerned.

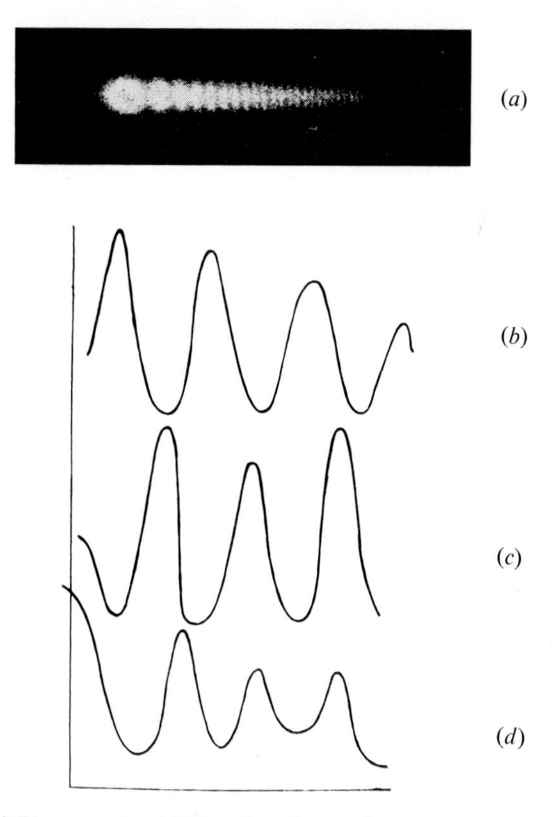

Fig. 2.23. (*a*) Photograph of " Lloyd's mirror " fringes using a laser source illuminating a slit; interference occurs between light emerging from the slit and that reflected from a sheet of glass. (*b*) Plot of intensity variation in Lloyd's mirror fringes produced with a microwave source ($\lambda = 3$ cm) and a wooden bench top as reflector. (*c*) Plot of intensity variation in Lloyd's mirror fringes produced with sound waves (frequency = 11 kHz, $\lambda = 3$ cm) using the wooden bench top as reflector. (*d*) Densitometer trace of the photograph of (*a*). The scales have been adjusted to give comparable sizes—the difference in location of the first maximum arises from the different phase shifts at the mirror.

Fig. 2.23 shows a comparison between so-called ' Lloyd's mirror ' fringes produced (*a*) optically, (*b*) with microwaves and (*c*) with ultrasonics. For comparison purposes we show also at (*d*) a microdensitometer trace of the pattern of (*a*); the departure from a cosine curve is due to the non-linearity of response of the photographic film.

Electron patterns may be recorded either with a photographic plate or by means of a positively charged collector which detects the arrival of negative charges. Other particles may be detected by their ionizing properties and indeed in the early days X-rays were detected that way. X-rays also affect a photographic plate but more recently the practice has swung round again and it is again common to record X-rays with Geiger or scintillation counters.

2.8. *How can we measure coherence?*

Now that we have studied the nature of radiation and have seen something of the way in which scattering or diffraction patterns can be observed or recorded, we can return to the problem hinted at in section 2.2, the measurement of coherence. This is a somewhat complex operation and for this reason the section has been starred and may be omitted at first reading.

* Let us consider spatial coherence to begin with. Most of us will have carried out some kind of double-slit experiment using either a sodium flame or a sodium lamp and will be aware that an initial ' source-slit ' is necessary between the flame and whatever device is being used to produce the two effective sources. We shall assume for the moment that Young's arrangement with two further slits is being used as in fig. 2.24, but the principles apply equally well to systems using a bi-prism, split-lens, etc.

What is the process by which the single source slit introduces spatial coherence? It is, in fact, itself a diffraction process.

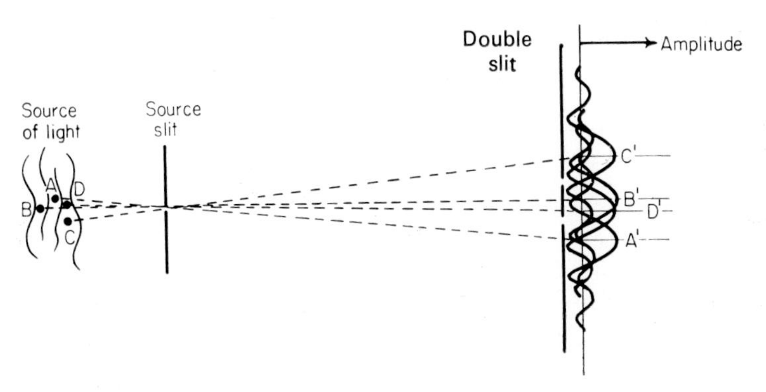

Fig. 2.24. Diagram showing how a single slit introduces some degree of spatial coherence in illuminating a pair of Young's slits.

Fig. 2.24 is intended to be a section of the system by a plane perpendicular both to the planes of the slits and screen and to the length of the slits. Consider one point of the source in this plane such as A. If this were the only source of illumination the single slit would produce a diffraction pattern whose cross-sectional amplitude distribution would be as shown at A′. (It would look something like fig. 2.18 without the fine fringes. Similarly a point such as B would produce a pattern at B′ and C at C′ etc. When the whole source is considered, the illumination of the double slit will be the superposition of a great many laterally displaced single-slit patterns, and for many of these the two elements of the double slit will both be within the same central peak. This is the origin of the coherence introduced by the single slit.

Now if the source-slit is *broad*, the diffraction pattern will have a narrow central peak and so the chance of many such peaks lying over the two slits is small and the coherence is low. If the source slit is *narrow*, its diffraction pattern has a broad central peak and many such peaks may span the two elements of the double slit and the coherence is high.

Notice that for certain single slit widths it is possible for the central peak to be over *one* element of the double slit and the first subsidiary maximum—which is in the opposite phase—to lie over the other. The number of elements of the source for which this would occur is small but nevertheless it does suggest that *negative* coherence might be possible. We shall return to this point a little later.

Now we must consider the experimental measurement of coherence. If the illumination of the double slit were totally coherent spatially, that is, if we had parallel wave fronts impinging on it, and if the two elements of the double slit transmitted identical amplitudes, then the interference pattern produced would have an intensity distribution which would vary sinusoidally. It would alternate from some maximum value to zero. If the light were totally incoherent spatially, then the sinusoidal distribution would disappear. It should not be too surprising to find, therefore, that for partially coherent light we have a mixture of the two—that is the elements of the source giving central peaks covering both slits give a sinusoidal distribution with real zeros and the elements not covering both slits give uniform illumination. The more highly coherent the light illuminating the two slits the nearer will the low points in the sinusoidal approach to zero and the contrast or ‘ visibility ’ of the fringes is a measure of the degree of coherence.

In mathematical terms the visibility function V is defined as $(I_{max}-I_{min}/I_{max}+I_{min})$, where I_{max} is the intensity of the central bright fringe and I_{min} that of the first minimum, and this quantity V has the same magnitude as the degree of coherence. Fig. 2.25 shows three patterns using the same pair of double slits for each. For (*a*) the source is very small and therefore the coherence is high; for (*b*) the source is larger, the coherence much lower, and in fact, has the *negative* correlation referred to above, as witnessed by the intensity *minimum* at the centre; for (*c*) it is larger still and the coherence is lower, but the correlation is again positive.

What about the measurement of temporal coherence? You will recall that in section 2.1 temporal coherence was described as being concerned with the phase relationships between displacements at different instants of

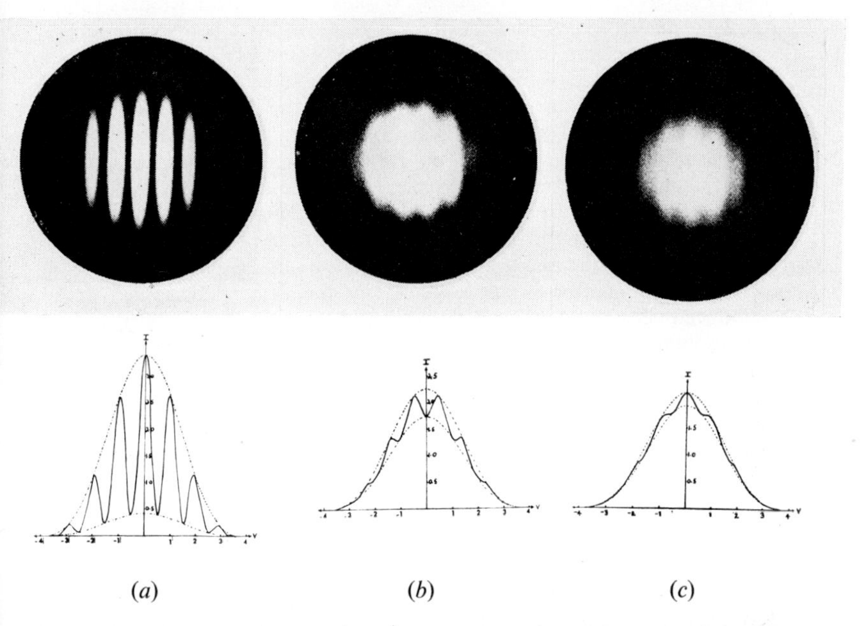

(*a*) (*b*) (*c*)

Fig. 2.25. (*a*) Diffraction pattern of two holes with, below, the theoretical intensity curve. (*b*) As for (*a*) but with illumination of lower coherence and showing negative correlation. (*c*) As for (*a*) but with illumination of even lower coherence showing restoration of positive correlation. By permission of Professor B. J. Thompson, Institute of Optics, Rochester, N.Y.

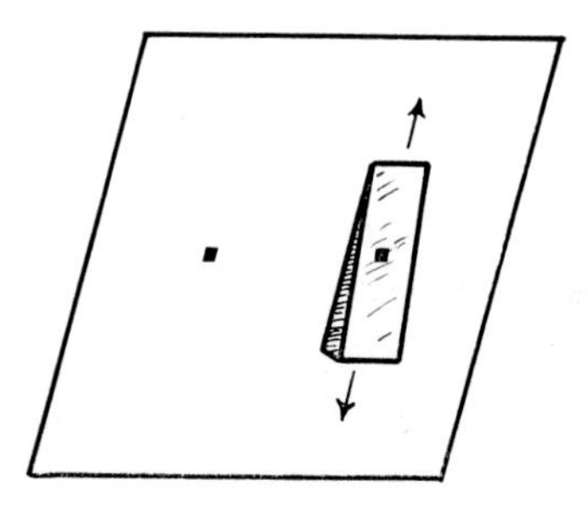

Fig. 2.26. Two apertures with a glass wedge over one which when placed in the parallel beam section of a diffractometer (e.g. at P in fig. 2.20 (*a*)) permits the measurement of temporal coherence.

time at a particular point in space. We also saw that the length of time for which a wave passed a particular point without a discontinuity—which is the measure of temporal coherence—could be translated into a coherence length simply by multiplying the time by the speed of travel of the waves. This coherence length forms the basis of methods of measuring temporal coherence which, in fact, closely resembles the technique for measuring spatial coherence. Fig. 2.26 shows a suitable experimental arrangement in principle. It is a two-slit diffraction experiment in which an additional optical path length can be introduced for the waves passing through one of the apertures. Ideally the apertures should be small holes and a wedge of glass or quartz is placed over one of them. By sliding the wedge to and fro, the additional optical path, introduced because of the reduction of the velocity of light in the wedge, can be increased until the visibility of the fringes, produced when the emergent waves interfere, disappears. The additional optical path is then a measure of the coherence length. Since one path is in air and the other in a medium of refractive index n (i.e. the velocity is reduced in the medium in the ratio $n : 1$) the additional path is $(n-1)t$, where t is the thickness of the wedge at the position of the hole.

2.9. *Images with self-luminous and incoherently illuminated objects*

In most of the detailed discussions of this book we confine attention to coherent illumination, but it is worthwhile to consider briefly the consequences of departure from coherence. If we consider scattering from any two points on the object, the resultant pattern is of sinusoidal fringes (see, for example, fig. 2.8). If the illumination is incoherent, or the two points are self-luminous, the pattern is still of sinusoidal fringes *instantaneously*, but the positions of the fringes continually changes as the relative phase of the waves radiated from the two points changes. The net result appears to be uniform illumination.

However, if we then proceed to the second stage of imaging—recombination—the instantaneous phase relationships are preserved and the image is formed without difficulty, regardless of the illumination. Thus for any process that involves recording of a scattering pattern, interpretation will only be possible with coherent illumination. But if we continue with recombination to produce an image the process is still possible whatever the illumination.

One final, and important, point is that (as we shall see in Chapter 5), in all practical imaging systems, the scattering pattern is limited, or modified, by apertures or defects with resulting degradations of the image. Detailed studies, that are beyond the scope of this book, show that the kind of degradation depends on the nature of the illumination and incoherent illumination can result in better resolution. (A hint towards an explanation of this is that, statistically, the rapid phase changes lead to more information entering the recombination stage).

3. Principles of direct recombination processes

"...... all concentrating, like rays
Into one focus kindled from above;"

Lord Byron
Don Juan, canto ii st 186

3.1. '*Straight line*' *imaging with pinhole cameras*

We must now return to the problem of how the second stage of the image-forming process can be achieved. Let us think back to the experiment in which we placed a slide in a projector with no projection lens and produced a patch of light on the screen: is there an even simpler way of producing an image than by replacing the lens? We said that the problem is just that *every* point on the screen receives information from *every* point on the slide and we are really being embarrassed by an excess of overlapping information. Is there any way in which we can, by sacrificing a great deal of the information, still leave ourselves with enough in a more manageable form to enable us to form an image? The answer is surely 'yes'. By placing a card between the slide and the screen and piercing one small hole in it we can ensure that light from each point on the slide falls on only a very small region of the screen and simple straight line geometry shows us that the arrangement of the patches on the screen will be exactly the same as that of the corresponding areas on the slide, except for inversion of the top and bottom and left and right (fig. 3.1). This, of course, is the principle of the pinhole camera. The size of the image is determined purely by the geometry and depends only on the distances between the components (fig. 3.2).

This system is so simple that it may be at first surprising that it is not more widely used. What are the snags? The first is that, if the hole used is large enough to produce a bright image, then the patches of information overlap and the result is blurred: if, on the other hand, the hole is made smaller then a sharper image results but, since more of the light is being cut out by the screen, the image is very dim. The

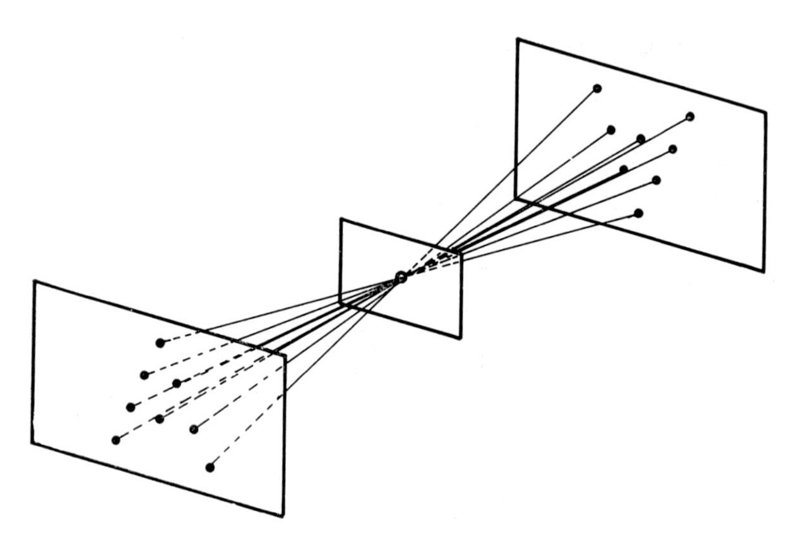

Fig. 3.1. Principle of the pinhole camera.

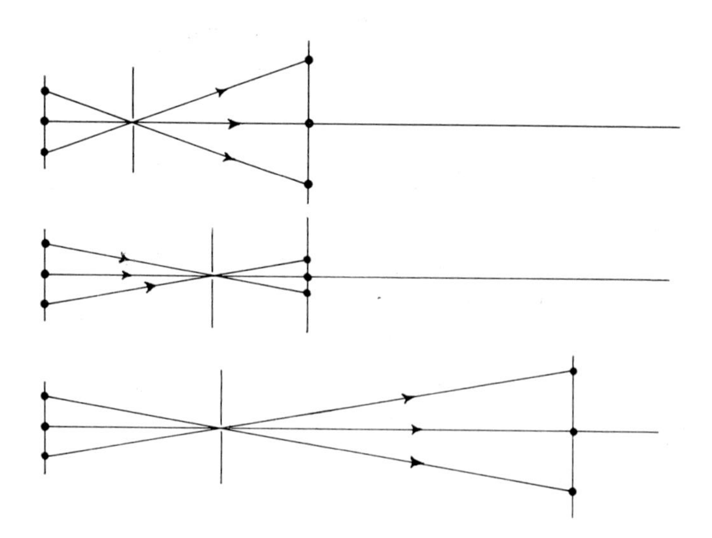

Fig. 3.2. The size of the image is determined by the ratio of the distances between components.

(*Opposite*). Fig. 3.3. (*a*) and (*e*) Objects photographed with a good lens. (*b*) and (*f*) Images with a large pinhole. (*c*) and (*g*) Images with a medium pinhole. (*d*) and (*h*) Images with a small pinhole.

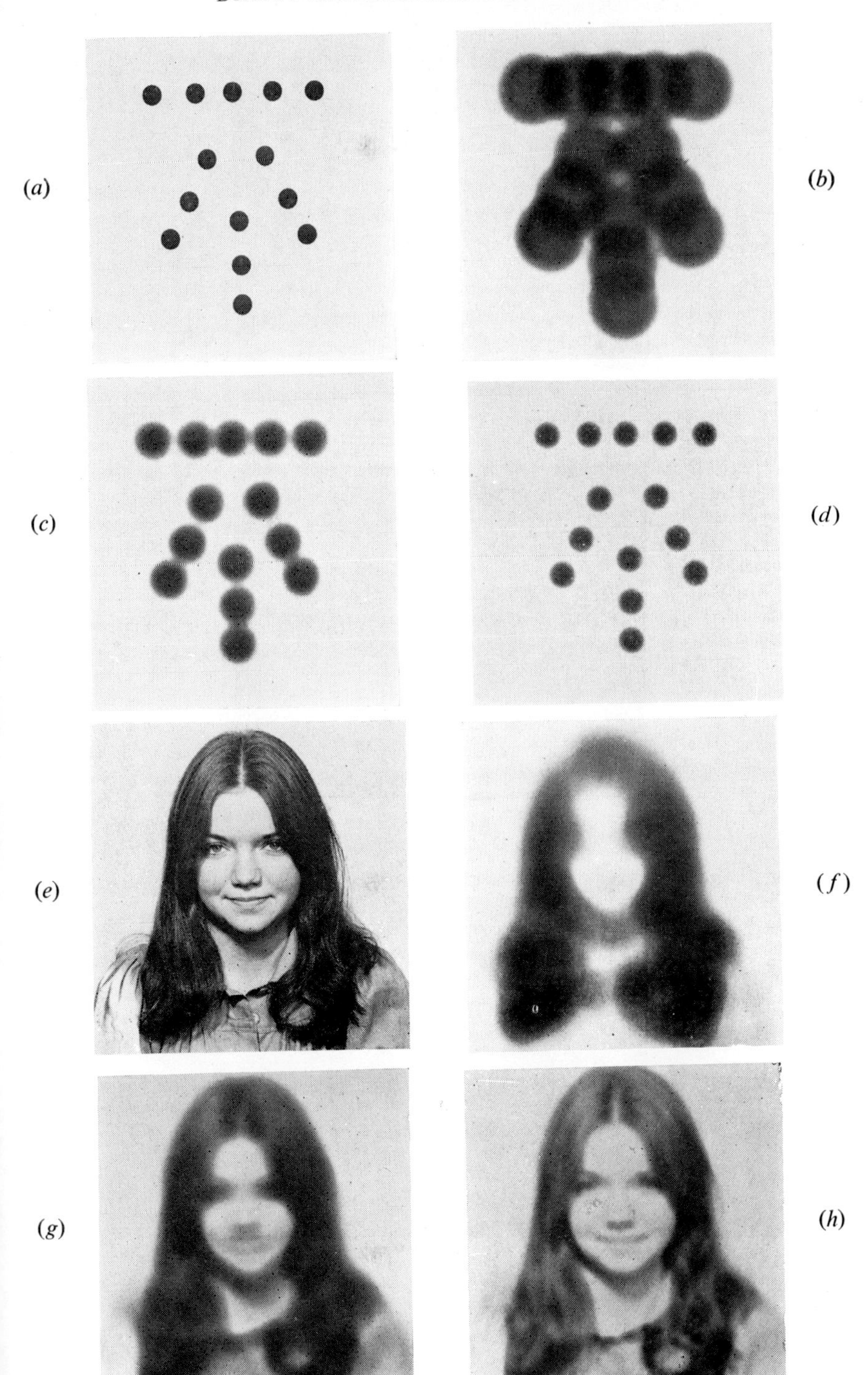
(a)
(b)
(c)
(d)
(e)
(f)
(g)
(h)

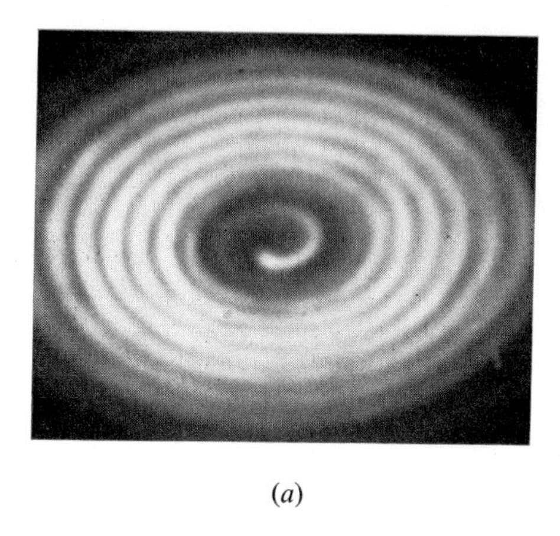

(*a*)

(*b*)

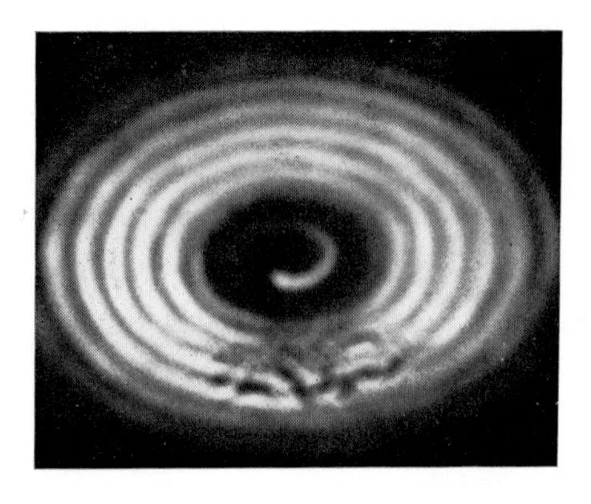

(*c*)

Fig. 3.4. (*cont. opposite*).

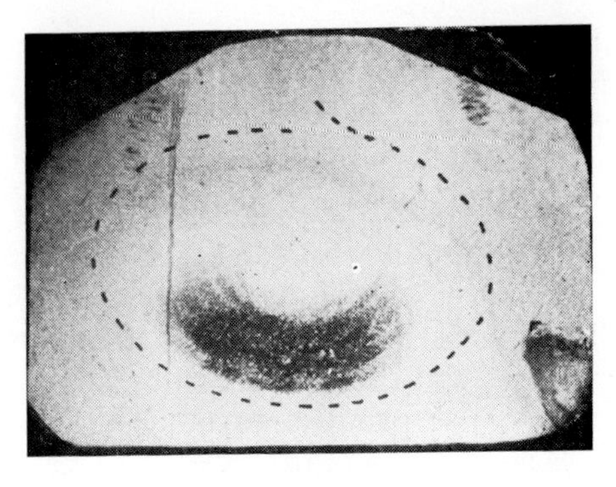

(*d*)

Fig. 3.4. (*a*), (*b*) and (*c*) X-ray pinhole photographs of target of X-ray tube showing progressive deterioration. (*d*) Photograph of target after tube had ceased to operate and had been dismantled. By permission of Mr. W. Sutherland, Velindre Hospital, Cardiff.

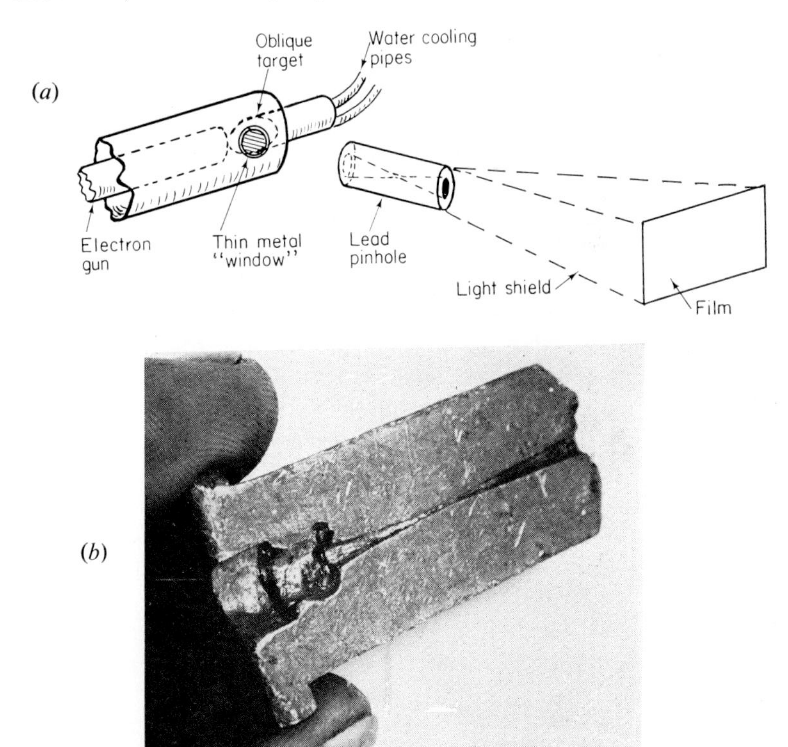

Fig. 3.5. (*a*) Experimental arrangement for producing fig. 3.4 (*a*)–(*c*). (*b*) Cross-section of lead ' pin-hole ' used for fig. 3.4 (*a*)–(*c*).

second is that if the hole is made really small in an attempt to obtain very sharp images the desired result does not follow. Apart from the fact that the image may become too dim to see, the hole produces its own diffraction pattern at each point of the image and, as the hole gets smaller, its diffraction pattern gets larger; the patches thus begin to spread again. In the limit when the hole becomes smaller than the wavelength of light the light emerges from the hole and is scattered uniformly in all directions and so we are back to the first stage of total overlap of information and with hardly any light anyway!

Fig. 3.3. shows pinhole photographs of two sorts of object—one made up of separate points, which enables us to see the effect of the pinhole size and the other a rather more complicated—and attractive—object; images produced with three sizes of pinhole are given together with a photograph using a good lens.

Although it must now be clear that this is not a very practical system of imaging there are some circumstances in which it has proved useful. For example the target of an X-ray tube cannot easily be examined for damage because it is permanently sealed inside a vacuum chamber and the ‘ window ’ through which the X-rays emerge is often a thin metal foil which is opaque to visible light. X-rays cannot be focused by material or electromagnetic lenses and hence, if we wish to use the X-rays emerging to image the target, the only direct way open to us is to use a pinhole camera. Fig. 3.4 shows a series of pinhole photographs taken some years ago by Mr. W. Sutherland of the Velindre Hospital, Cardiff, showing progressive deterioration of a target under electron bombardment. It is interesting to note that here the target is being bombarded by electrons and is emitting X-rays so that the object is really ‘ self-luminous ’. The final picture is a normal photograph taken when the tube finally ceased to work and was taken apart. Fig. 3.5 (*a*) is a diagram of the experimental arrangement and fig. 3.5 (*b*) is a photograph of a section of the ‘ pinhole ’ used which, of course, has to be a massive piece of lead to provide the necessary opacity to the X-rays. A further point of interest is that the electrons are emitted from a hot spiral filament and, under the influence of the high voltage, they travel in straight lines; it is an emission microscope (next section) but with unit magnification. The spiral ‘ image ’ of the filament on the target can clearly be seen in the photographs.

3.2. *Field emission microscopes*

Because of the simple geometry by which they produce images, two microscopes will be described. The first is the true field emission microscope and the second is the closely related field-ion microscope.

In the field emission microscope the radiation used in the imaging process is electrons and it turns out if one examines the consequences of certain aspects of quantum mechanics that it is possible for a cold surface to emit electrons provided that a potential gradient of the order of 5×10^9 volt per metre is produced. On the face of it this is an enormous value but it turns out that if the specimen is small enough it can be achieved. The specimen is a metal wire the end of which has been polished (usually by electrolytic means) to a fine point which may be much less than one micrometre in diameter. This point is placed in a vacuum chamber and there is a fluorescent screen perhaps 10 cm or so away from the point. The application of a potential difference of only a few kilovolts between the point and the screen can create the necessary potential gradient close to the point of the wire. In practice the wire is usually heated and, since the chamber is evacuated, any electron emitted is accelerated very rapidly by the field away from the point of the wire. It will travel in a straight line and of course create a glow when it hits the screen. The emission of radiation from the tip of the wire will depend very much on the atomic distribution at the tip and the result is that a magnified reproduction of the atomic geometry of the surface of the wire is produced on the screen. If the tip (fig. 3.6 (*a*)) is assumed to be spherical and of radius r, then the effective magnification, since the electrons are emitted normally from the surface and behave as though emitted from the centre of the hemisphere, is simply the ratio of the distance from the tip to the screen divided by the radius. If the screen distance is 0·01 metre and the radius of the tip is 0·01 micrometre, a magnification of a million is achieved.

Fig. 3.6 shows an optical analogue of the principle. An extremely small bright source of light was set up about 15 metres from a screen. The object, consisting of a piece of twin flex, was placed successively nearer to the source; the magnification achieved by this ‘ straight-line ’ imaging is obvious but it is also clear that the wavelength relationship to the size of the object is having an effect and, in 3.6 (*c*) particularly, very strong evidence of Fresnel diffraction effects can be seen.

It can be seen that this system very closely resembles the pinhole camera in its geometry and mode of operation. The other microscope mentioned above—the field ion microscope—operates very much on the same principle except that here the imaging radiation is positive ions. The geometry of the system is very much the same but the wire is charged positively and a trace of gas (usually helium) is admitted to the chamber. Ionization occurs and it is the helium ions which are accelerated away from the tip to produce the image.

(*a*)

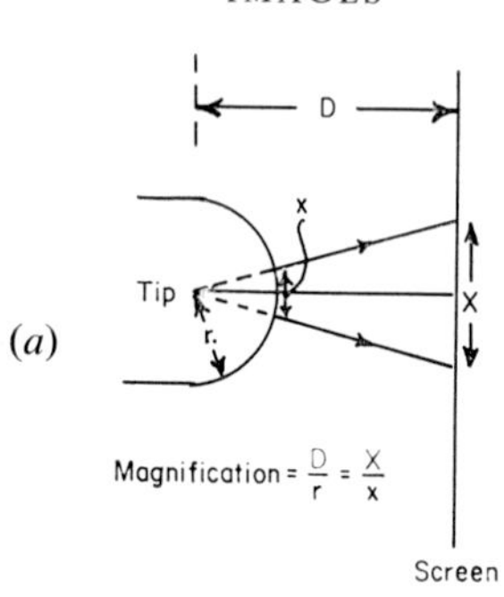

(*b*)

(*c*)

Fig. 3.6 (*cont. opposite*).

(*d*)

Fig. 3.6. (*a*) Geometry of field-emission and field-ion microscopes. (*b*), (*c*) and (*d*) Optical analogue showing how large magnification can be produced if an object is placed very close to a tiny source of light. The pictures also show Fresnel diffraction fringes round the object.

The question might be asked why the ion microscope has any advantage over the field emission microscope and the answer is tied up with the resolution. In order to obtain a really good resolution it is essential that the pattern of electron or ion beams leaving the surface of the hemispherical end of the wire remains completely unchanged except for size in transit to the screen. It turns out in practice that random thermal motion of the atoms in the tip is the biggest limitation and tends to produce fuzziness in the resulting image. In order to avoid this problem the specimen needs to be cooled to as low a temperature as can conveniently be obtained and of course this in itself makes the electron emission process somewhat difficult. The result is that the field ion microscope with the tip of the wire cooled by liquid nitrogen or some other coolant produces the best resolution and, in fact, is capable of revealing details down to about 2×10^{-10} m. This is quite sufficient to reveal the patterns of atoms, for example, in most metal structures. Fig. 3.7 shows a field-emission micrograph and a field-ion micrograph.

3.3. *Lenses for light*

In order to demonstrate the transition between the pinhole and the lens as image-forming systems, there is a well known experiment (fig. 3.8) in which several pinholes are used in the card to increase the amount of light transmitted while retaining the clarity of the image associated

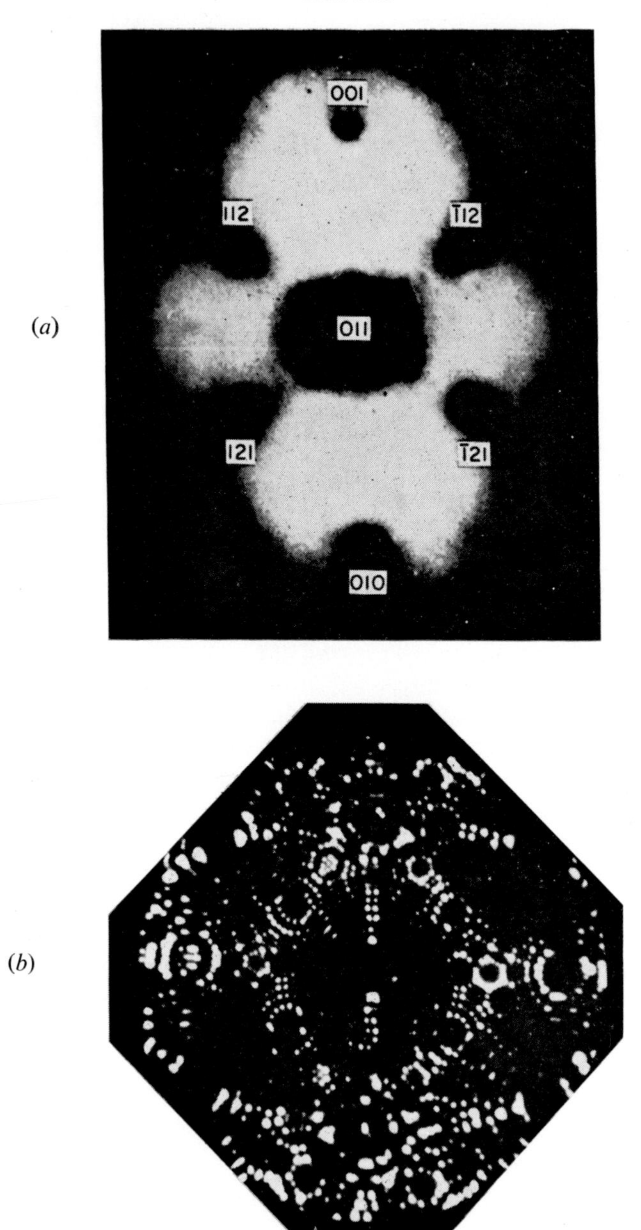

Fig. 3.7. (*a*) Field-emission micrograph from tungsten surface; the labels refer to crystallographic directions. (*b*) Field-ion micrograph from tungsten surface. (From *Introduction to Modern Microscopy*, H. N. Southworth, Wykeham Publications, 1975.)

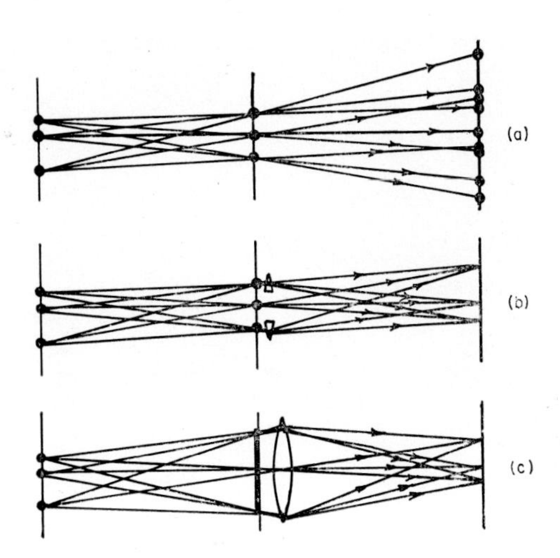

Fig. 3.8. Geometry of demonstration of transition from pinhole to lens. (*a*) 3 pinholes. (*b*) 3 pin-holes with prisms to deviate beams. (*c*) 3 pin-holes with lens.

with the small hole; the images (fig. 3.9 (*a*)) are not then coincident but can be made so by placing a prism of a given angle and in the right orientation over each hole (fig. 3.9 (*b*)). Investigations soon show that, as the number of holes is increased, each with its appropriate prism, the necessary prisms turn out to be effectively sections of a single lens. If the number of holes is increased until they merge with each other and the prisms are all replaced by a single lens the maximum use is made of all the light falling on the area covered by the lens while enhancing the sharpness of the image associated with the original pinhole (fig. 3.9 (*c*)). The clarity of the image of fig. 3.9 (*b*) must depend on the precision with which the prisms are made and adjusted and this requirement is carried through when a single lens is used: the optical precision of the shape of the lens surfaces is the primary factor governing the sharpness. One of the limitations on the sharpness of the image which were discussed for the pinhole camera still applies, namely the effect of the pinhole being too small. The effect is paralleled by the whole lens aperture being so small that its rim introduces noticeable diffraction effects itself. This aspect of lens behaviour is treated in section 5.1.

The approach to the operation of a lens via the pinhole camera is acceptable as a starting point but is not adequate to enable us to explain all the phenomena associated with lenses. A more valuable way of thinking about the operation of the lens involves consideration

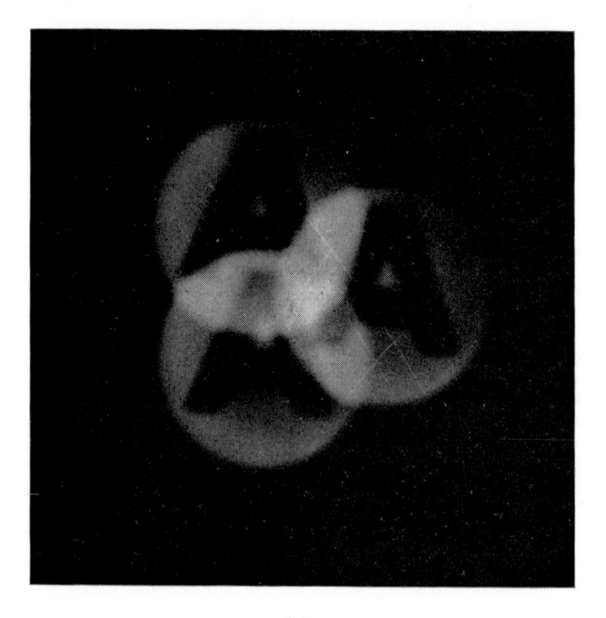

(*a*)

(*b*)

Fig. 3.9 (*cont. opposite*).

(*c*)

Fig. 3.9. (*a*) Photograph taken with system of fig. 3.8 (*a*). (*b*) Photograph taken with system of fig. 3.8 (*b*). (*c*) Photograph taken with system of fig. 3.8 (*c*).

of the phase changes which it introduces into the patterns of scattered waves.

In fig. 3.10 we have chosen one point on the object and the corresponding point on the image produced by the lens. What the lens does, in effect, is to ensure that by whatever path a wave travels from P to P′ the time taken will be identical; in other words, where the air path is larger (e.g. PAP′) the glass path is shorter and vice versa (e.g. PBP′). Since light travels more slowly in the glass the two effects can be used to compensate precisely. If the time taken is identical then all the parts of the wave arrive in phase at P′ regardless of the nature of the illumination. Clearly there can be no other point on the screen at which *all* the waves from P will arrive in phase and so we build up a one-to-one correspondence between points on the image and points on the screen—just as we did for the single small pinhole.

It may come as something of a surprise to realize that the resulting image is, in my terminology, the result of interference effects between the waves as rearranged by the lens. We have stressed that the waves arriving at P′ will all be in phase and so will add up to reproduce the image of P. Why is none of the light scattered by P reaching any

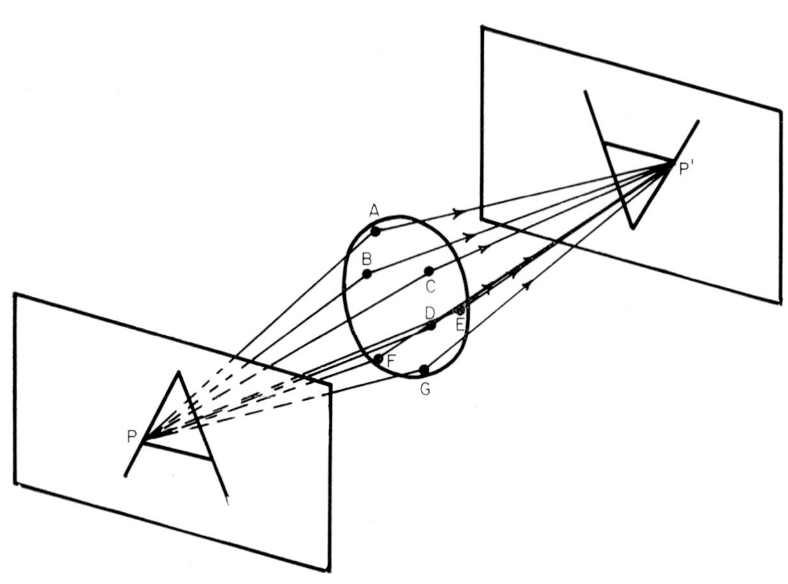

Fig. 3.10. A lens ensures that the time taken by the radiation to travel by any route such as PAP′, PBP′ etc. is the same for a given pair of points P and P′.

other point but P′? To answer this question it is best to consider P to be a single hole in an opaque screen. A perfect lens will then produce what is essentially a single bright spot on the screen and no light anywhere else. What happens, of course, is that at P′ all the waves are in phase and so we get an interference maximum and *everywhere else* the phase relationships are so random that the waves effectively cancel each other out and we have interference minima. If you have difficulty in understanding this last point, think about all the possible phases of waves arriving at any other point of the screen in the following way. If all possible phases exist in the collection we can always divide them up into pairs which differ by π radians, and each pair (e.g. $\frac{\pi}{4}+\frac{5\pi}{4}, \frac{\pi}{3}+\frac{4\pi}{3}$ etc.) gives zero resultant. This will not be true in the vicinity of the image point and this complication will be discussed later (section 5.1). The same argument can now be applied to each point on the object.

Thus the remarkable function that the lens performs is to ensure that all the phases of the waves falling on it are adjusted in such a way that this one-to-one correspondence arises. The fact that lenses are so common and that they are relatively easy to make tends rather to blind us to the precision and elegance of their operation.

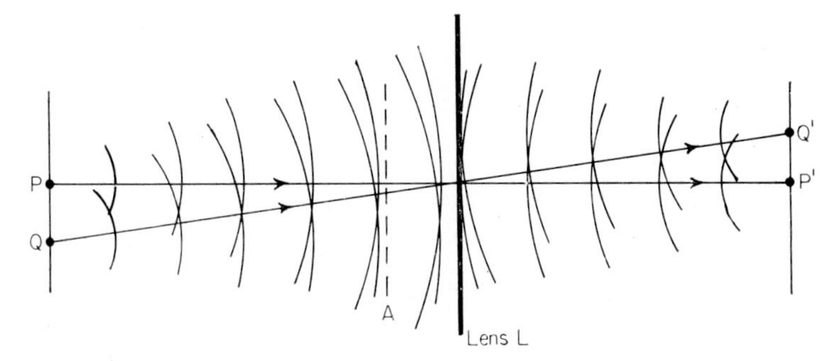

Fig. 3.11. The lens as a phase adjuster.

We should develop the idea of the lens as a phase-adjuster one stage further before leaving it. It is important always to bear in mind that we are dealing with electromagnetic waves in three dimensions. Let us first consider what actually happens when a single point source is imaged. Fig. 3.11 is a sectional view in which the function of the lens as a phase adjuster changing the curvature of the wave fronts emitted by P becomes clear. It performs this function for every point being imaged and it is only when one considers the combined effect of the waves emitted from all points of the image plane to produce an extremely complicated resultant set of waves at A, that one begins to recognize the remarkable nature of the action of the lens. The technique of considering the wave fronts at A phase-modified by the distribution of delaying material (glass in this case) in the plane L is a very powerful one which when translated into mathematical terms forms the basis of modern imaging theory. (We shall say a little more about this in section 5.4.)

* The behaviour of a lens as a phase-adjuster can be related to its behaviour as described by the conventional formula of geometrical optics in the following way:

In fig. 3.12 (*a*), consider the spherical waves diverging from the point P and falling on a plano-convex lens whose focal length f is equal to the distance from P to the lens. The lens has been *drawn* with a large thickness so that we can see what is going on inside it, but it is to be regarded as a thin lens. We have also drawn the angle OPA large but, as in geometrical optics, our result will only apply to paths making small angles with the axis. Let n_1 be the refractive index of the air, n_2 the refractive index of the glass and t the thickness of the lens on its axis.

We know that, by definition, the wave fronts must become plane parallel surfaces normal to the optical axis to the right of the lens since P is at the focal point on the axis. Consequently the optical path

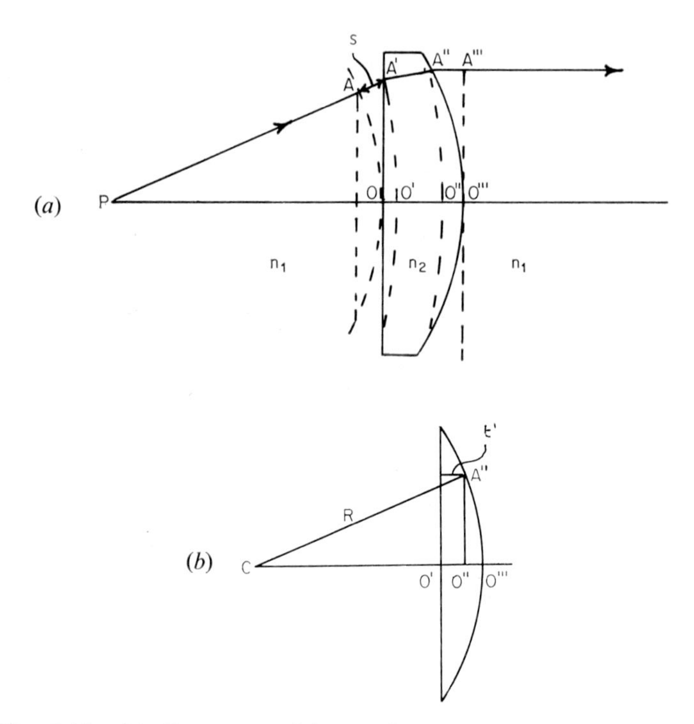

Fig. 3.12. (*a*) Geometry of lens acting as phase adjuster converting a spherical wave radiating from the back focal point P into a plane wave front O‴A‴. (*b*) Geometrical relationships for a spherical surface.

POO′O″O‴ must be identical with the optical path PAA′A″A‴ if the phase-adjustment is to produce these parallel wave fronts.

Along the centre line the total optical path is

$$n_1\mathrm{PO}+n_2\mathrm{OO}'''$$
$$=n_1f+n_2t$$

Along the path PAA′A″A‴ the total optical path is

$$n_1\mathrm{PA}+n_1\mathrm{AA}'+n_2\mathrm{A}'\mathrm{A}''+n_1\mathrm{A}''\mathrm{A}'''$$
$$=n_1f+n_1s+n_2t'+n_1(t-t')$$

where s is AA′, t' is the lens thickness A′A″ which, since (as already stated) the lens is supposed to be thin and the paths at only small angles to the axis, may be taken as the thickness measured parallel with the axis, i.e. to be equal to OO″.

These total optical paths must be equal. That is

$$n_1f+n_2t=n_1f+n_1s+n_2t'+n_1t-n_1t'$$
$$(n_2-n_1)\,(t-t')=n_1s.$$

Now let us apply Pythagoras's theorem to the triangle POA′ and make the substitution $OA' = h$. Then

$$(PA')^2 = (PO)^2 + (OA')^2$$

and hence

$$(f+s)^2 = f^2 + h^2;$$
$$f^2 + 2fs + s^2 = f^2 + h^2.$$

Now if our restriction to paths making small angles with the axis is again recalled, it is clear that s will be very small compared with f and hence its square may be neglected and we find

$$s = h^2/2f.$$

So the equality of optical path lengths is subject to the condition that

$$(n_2 - n_1)(t - t') = n_1 h^2/2f.$$

Now let us suppose that the lens surface is spherical—as it would be in the case of the thin lens studied in geometrical optics—and consider fig. 3.12 (*b*).

C is the centre of curvature of the lens surface and R its radius. We apply Pythagoras's theorem to the triangle CA″O″ and obtain

$$(CA'')^2 = (CO'')^2 + (A''O'')^2$$

which, again subject to the conditions already specified, is approximately replaced by

$$R^2 = [R - (t - t')]^2 + h^2$$
$$R^2 = R^2 + (t - t')^2 - 2R(t - t') + h^2$$

and again we can neglect $(t - t')^2$ as both t and t' are likely to be small compared with the other quantities and so we arrive at

$$t - t' = h^2/2R,$$

and the condition becomes

$$(n_2 - n_1)h^2/2R = n_1 h^2/2f$$

or

$$(n_2 - n_1)/R = n/f.$$

You may recognize this as the formula given by the more conventional geometrical optics approach using the same kind of approximations.

3.4. *A hybrid technique—the zone plate*

The pinhole camera was introduced as a means of imaging by discarding a large part of the information that falls on the card leaving, in effect, just one set of data which is sufficient to produce a decipherable image but lacks intensity. The lens was introduced as a means of using all the information on the card by rearranging the phases so that all the light waves cooperate to produce one-to-one correspondence between object and image. There also exists an intermediate technique which involves discarding about half the information and retaining the rest in such a way that it cooperates to produce a relatively bright image. This is the so-called zone plate. It is extremely interesting as an elegant example of optical theory (and is useful as a basis for explaining the ideas of holography in section 4.3); only recently have important practical applications in image formation been developed largely because of manufacturing difficulties which—as we shall see later—the technique of holography has itself provided a means of overcoming.

Let us first consider the problem of imaging a single point. We saw that the lens adjusts the optical length of all possible paths between the object and image points to be identical by inserting dielectric material of varying thickness. In order to achieve constructive reinforcement it is not necessary, however, for the optical path lengths to be *identical.* If the optical paths differ only by multiples of a wavelength of the radiation concerned the phases will differ only by multiples of 2π: there is no *practical* difference between phase differences of $0, 2\pi, 4\pi$, etc. (Provided that the temporal coherence is sufficiently high, i.e. the coherence length is great enough.) In fig. 3.13 the paths POP′ and PAP′ differ by 1 wavelength, the paths PAP′ and PBP′ differ by 1 wavelength and so on. Thus waves travelling by all these routes arrive effectively in phase at P′ just as though a lens had been placed in the plane OE.

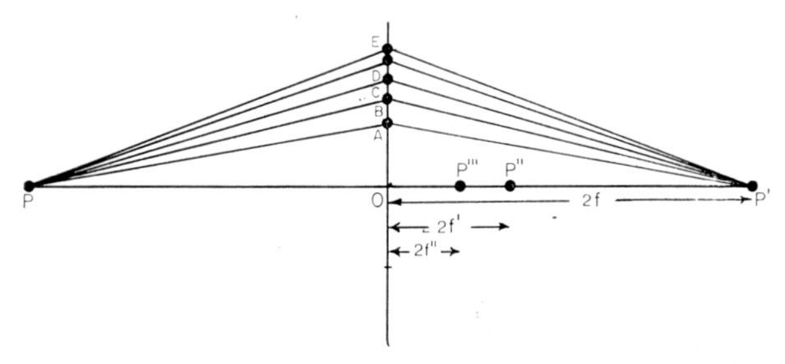

Fig. 3.13. Geometry of a zone-plate. Paths PAP′ and PBP′ differ in length by one wavelength.

Thus, if we devise a mask which effectively cuts out those parts of the wavefront which would travel via paths intermediate between POP′ and PAP′ etc. we shall have achieved our aim of eliminating some of the information and leaving the rest in an interpretable form. As described so far, our screen would consist of a series of very thin annular apertures and very little light would be transmitted. Fresnel showed that, in fact, the annuli can be made very much wider, provided that certain conditions are obeyed, without upsetting the optical behaviour, and hence the amount of light transmitted can be quite large. The full theory is rather beyond the scope of this book but is based on the notion that if we make each annulus of such a width that the change in path length between waves travelling via the inner edge and via the outer edge is half a wavelength the focusing effect is the same as for the very narrow annuli. But we get 50 per cent of the light transmitted instead of a tiny fraction. The theory is known as the Fresnel ‘ half-period zone ’ theory and is given in detail, for example, in *Optics* by Smith and Thomson. Fig. 3.14 is a photograph of a Fresnel half-period zone plate which will act as a lens and fig. 3.15 is a photograph taken with the camera lens replaced by a photograph of fig. 3.14 on 35 mm film in which the overall diameter was 0·02 m.

Further consideration of fig. 3.13 will show that a zone plate with narrow annuli will not have just *one* ‘ focal length ’. There will for

Fig. 3.14. Fresnel half-period zone plate.

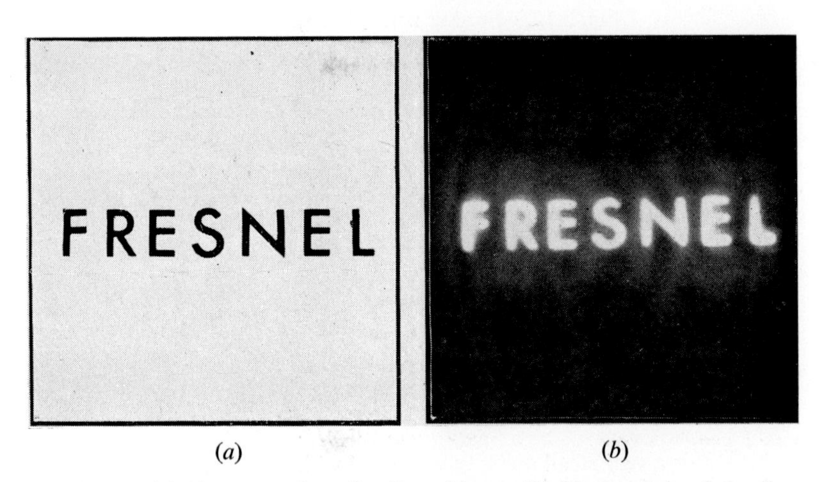

(*a*) (*b*)

Fig. 3.15. (*a*) Contact print of a line object. (*b*) Photograph of the line object using the zone plate of (*a*) in place of the camera lens.

example be a point P″ where all paths differ by 2λ, a point P‴ where they differ by 3λ and so on. There will also be points $-P'$, $-P''$, etc. on the other side of the zone plate which can be regarded as 'virtual' foci, since any set of waves which diverge from the zone plate as though from these points will have the same phase-difference relationships as those which come to a real focus at P′, P″, etc. It can be shown that a Fresnel half-period zone plate also has the same additional foci. The point about the virtual focus is often omitted in studies of the zone plate but it is useful to introduce it here since we shall need to use the idea again in talking about the principles of holography.

Although the full half-period theory is difficult it may be worth applying simple geometry to the given facts of the Fresnel zone plate to discover how the dimensions relate to the focal lengths.

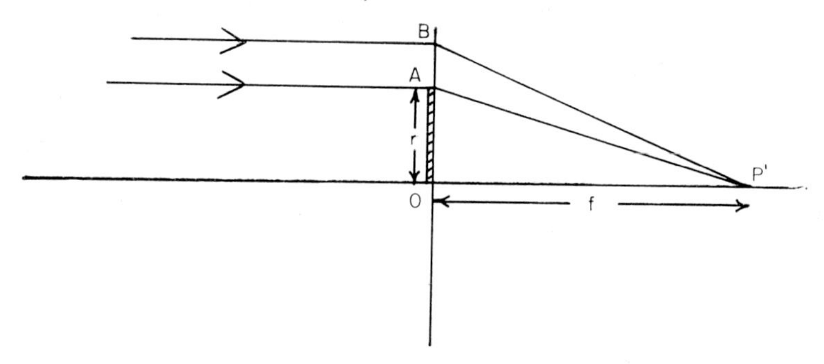

Fig. 3.16. Further zone-plate geometry.

* To keep the geometry simple we will consider parallel plane wave fronts coming in from the left in fig. 3.16 and being 'focused' at P′, one of the real focal points of the zone plate OB.

Since this is a half-period zone plate the central opaque zone of radius r must be such that the optical path AP′ is half a wavelength greater than the optical path OP′ (assuming that P′ is the most distant of the focal points).

Using Pythagoras's theorem we can see that

$$(AP')^2 = r^2 + f^2$$

or

$$(f + \lambda/2)^2 = r^2 + f^2$$
$$f^2 + \lambda f + \lambda^2/4 = r^2 + f^2.$$

Now λ^2 is a very small quantity compared with everything else so we can ignore it and write

$$f = r^2/\lambda, \quad \text{or} \quad r = \sqrt{(f\lambda)}.$$

Here r is the radius of the inner edge of the first half period zone. Following the same kind of reasoning we could find the radius of the inner edge of the pth half period zone r_p for which

$$(f+p\lambda/2)^2=r_p^2+f^2$$
$$f^2+p\lambda f+p^2\lambda^2/4=r_p^2+f^2$$

and, using the same approximation, we arrive at

$$f=r_p^2/p\lambda, \quad \text{or} \quad r_p=\sqrt{(pf\lambda)}.$$

Finally we could extend the reasoning to other focal lengths. For example the next shortest focal length f' (fig. 3.13) would involve path differences of $3\lambda/2$ instead of $\lambda/2$ and so the path-difference equation would become

$$(f'+3p\lambda/2)^2=r_p^2+f'^2$$

which leads to

$$f'=r_p^2/3p\lambda, \text{ or } r_p=\sqrt{(3pf'\lambda)}.$$

Similarly f'', the next focal length would be $r_p^2/5p\lambda$ and so on. So $f=3f'=5f''$.

For the zone plate used for fig. 3.15 $r_1=6{\cdot}7\times10^{-4}$ m and $f_1=0{\cdot}7$ m (for He–Ne laser light of wavelength $6{\cdot}33\times10^{-7}$ m). The zone plate can be thought of as a kind of phase-adjuster but it operates on the rather brutal principle of rejecting everything that is not of the right phase for its purposes rather than, as the lens does, by actually changing the phase to suit. It is not too surprising however to find that it obeys the normal lens formula of geometrical optics for any of its values of f.

There is another type of zone plate which has recently assumed considerable importance in practical optical systems. By substituting varying thicknesses of transparent material to alter the phase of the light passing through instead of opaque and transparent strips, and by making the variation in thickness continuous, rather than like a step function, it proves to be possible to produce a zone plate with only one real and one virtual focal point. It also uses practically all the light falling on it. In other words, it behaves exactly like a lens, but can be extremely thin even if the aperture is very large. In systems in which scatter from minute specks in the glass or absorption in the thickness of the glass is critical, these new zone-plate lenses may prove to be of great value. The problem has been of how to manufacture them. A technique based on holography has now been developed and will be described in Chapter 6.

3.5. *Lenses for electrons*

One of the most powerful image-forming devices is the electron microscope and this conforms to the general principles outlined so far.

Electrons are scattered by the object and the resulting pattern needs to be recombined or decoded. Electrons are very rapidly absorbed by most solid materials, so that the production of lenses comparable with those for visible light is not very practical. Fortunately electrons, being charged particles, are affected by both electric and magnetic fields. As a result electron ' lenses ' may be produced. Their actual operation is highly complicated but, fortunately, we find in practice that, if the aperture is kept very small, their behaviour can be regarded as phase adjusters that ensure a one-to-one correspondence between points on the object and image planes—though in the case of the electron lenses the inversion effect is replaced by a rotation of the whole image which varies with the object and image distances. This is a minor point however as far as we are concerned in this particular study. Fig. 3.17 shows the actual paths taken by electrons through an electromagnetic lens and it is the helical nature of these paths that gives rise to the rotation of the image.

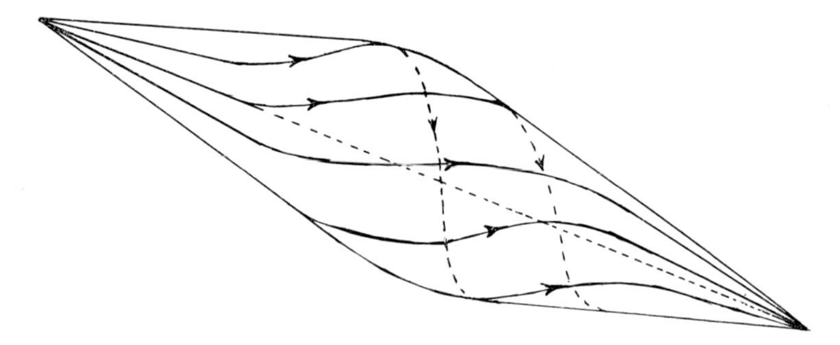

Fig. 3.17. Helical paths of electrons in electron-microscope lens.

From the many practical problems that arise in translating what appears to be a very simple device in principle into the very powerful tool that the electron microscope has become, we shall discuss only three. The first is that the object being studied needs to be inside a high-vacuum system and this immediately rules out many possible objects. Great ingenuity has been employed in finding solutions to this particular problem, for example, by preparing cast replicas of delicate objects that are then robust enough to be placed in the vacuum system; these are, however, outside our main concern and are discussed in detail in the Wykeham monograph No. 33 entitled *Electron Microscopy and Analysis* by P. J. Goodhew.

The second problem is that, as with optical microscopy, the image lies in a plane even though the object may be three-dimensional.

The one-to-one correspondence which we have come to recognize as the hallmark of a lens system relates points in two planes and hence only one plane of the object can be properly imaged; the others will be ' out of focus '. This effect—termed ' depth of focus '—is familiar to photographers and indeed arises whenever a lens system is used in the conventional way.

The third problem that must be mentioned here is that of lens perfection. Unfortunately—largely because of limitations on the precision of mechanical machining operations—electromagnetic lenses of large aperture are extremely difficult to make. In most practical electron microscopes the effective aperture is very small indeed relative to that possible with optical microscopes. Consequently, although electrons have the right kind of wavelength for imaging atoms, the practical apertures available restrict resolution to relatively large groupings of atoms.

The problems of depth of focus and of resolution limits are both discussed in more detail in Chapter 5 and some solutions are described.

3.6. *Imaging by scanning*

We should now remind ourselves once more of the basic problem that is our main concern in this chapter. In the first stage of the image-forming process radiation is scattered or emitted by the object and every point on the receiving plane contains information about every point on the object. Our problem is to disentangle the information. The pinhole camera achieved this at each point of the image by eliminating the information about all other points on the object and the consequent disadvantage is that so much of the scattered radiation is discarded that the resulting image is comparatively weak. The lens achieves a much more satisfactory result by rearranging the information so that a high proportion of the radiation scattered by one point arrives at a particular point on the image. But suppose we are dealing with radiation for which no lenses exist and we seek a degree of magnification and a level of illumination that rules out the pinhole system. Is there any other approach we could adopt? Suppose that instead of eliminating the information about every point but one we adopt the opposite plan. We choose a particular point on the object and extract from every point on the image plane the information about this particular point. (A lens, of course, does just this, but it does it for all points *simultaneously*; here we propose to take each point in turn.) We should not be throwing information away and so would not have the disadvantages of the pinhole camera. We should however need to take the object point by point in order to build up the whole image and this would mean that instead of instantaneous imaging the process would take time. Fortunately this is very simple to achieve—at least in theory. In figs. 3.18 (*a*) and (*b*) we start with a pinhole camera and simply move the pinhole until it touches the object. The pinhole can then be moved around on the object and for each point we shall obtain a screen-full of scattered radiation. But now the screen will be uniformly illuminated and, at any one position, we shall have information relating only to P_1, or to P_2, or to P_3 etc. (In fig. 3.18 (*b*) P_3 is the point selected.) Since the screen is uniformly illuminated we can place a single detector, e.g. a photo-cell, anywhere on it (for example at C in fig. 3.18 (*b*)). As we move the pinhole from point to point P_1, P_2, P_3 etc., the cell C will produce a signal corresponding to the information relating to each point.

We now wish to build up an image of the original object and this can be done with a replica of the moving pinhole system in which the pinhole might be replaced by a small lamp whose brightness is controlled by the output of the cell C (also shown in fig. 3.18 (*b*)). Thus when the lamp is at position L_1 it will reproduce an intensity corresponding to the light scattered by P_1 and so on. In this way the complete one-to-one correspondence can be built up in the time taken for the complete coverage of the object by the pinhole.

This is one example of the process called scanning; in practice it may be achieved in many different ways, some of which involve moving a spot of illuminating radiation systematically over the object and others the dividing up of the scattered radiation. It has one additional advantage that makes it of great practical significance and that is that it converts two- or even three-dimensional information about the object into a sequence of signals which vary in *time* (for example in fig. 3.18 (*b*) this would be the signal from the cell C controlling the lamp L). This of course is essential for transmission of information over radio or telephone links and for many other purposes where the simultaneous transmission of vast amounts of data would be difficult or impossible. It thus lies at the heart of all systems of recording and transmitting images via electronic apparatus.

One of the earliest applications was in television when the famous Nipkow disc (fig. 3.18 (*c*)) provided an effective means of transmission. It consisted of a spiral of holes pierced in a rotating disc that behave exactly as the pinhole in fig. 3.18 (*b*), traversing the object line by line. The receiver consisted of an identical disc moving over a neon lamp whose brightness was controlled by the incoming signal. Thus instead of moving the lamp, as in fig. 3.18 (*b*), a large lamp was scanned by the holes.

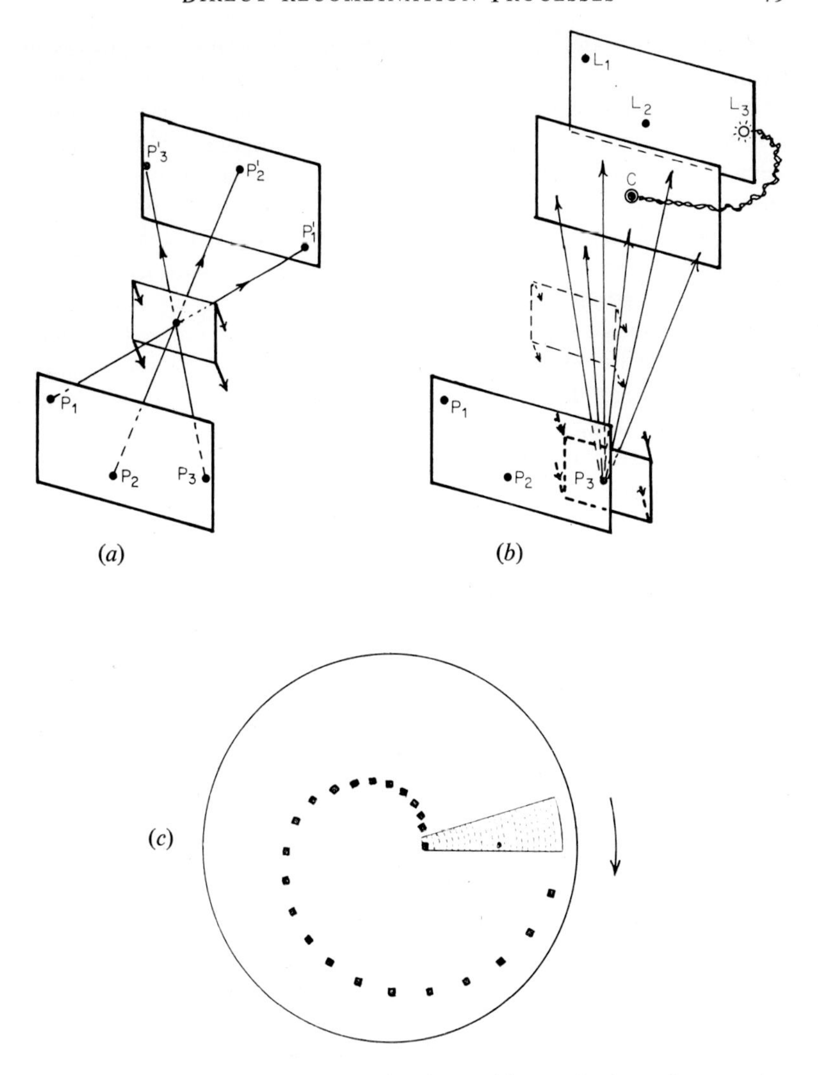

Fig. 3.18. (*a*) Pin-hole camera can be changed into point-by-point scanning system by moving pin-hole towards object. (*b*) When the pinhole is in contact with the object, only one point of the object can be studied at once and radiation from it covers the whole screen. A photo-cell at C would thus respond to the amount of scattering from P_1, P_2, P_3 etc. as the pinhole is scanned across the object. If the lamp is moved in synchronism with the pinhole to L_1, L_2, L_3 etc. and its brightness is controlled by the output of C the lamp will produce an image of the object point by point. (*c*) The original Nipkow disc used for scanning in early television. In this case a spot of light was made to scan across the object: the result is the same as if the pinhole had scanned across in contact with the object.

In practice, modern television cameras use much more sophisticated electronic scanning systems, but the principle of taking the information point by point remains identical. Similarly in radar and ultrasonic systems and in certain kinds of electron microscopes scanning techniques are used. Very often the object is illuminated by a narrow beam of radiation so that at any one moment only one point on the object is scattering. If you think about it you will see that the principle is exactly the same as that already described. We shall discuss some specific applications in detail in Chapter 6. Imaging by scanning is not quite a direct recombination process since the image has to be built up point by point. It has been included in this chapter because it finds a place in devices such as television cameras which scan with such rapidity that, to a human observer, the picture appears to be transmitted instantaneously.

In the next chapter on indirect recombination we confine our attention to techniques that underline the two distinct steps in the image-forming process by requiring the observer to do two separate experiments in order to achieve first scattering and then recombination.

4. Principles of indirect recombination processes

" More ways of killing a cat than choking her with cream "
Charles Kingsley
Westward Ho, ch. 20

4.1. *The essential problem in X-ray, neutron and electron diffraction*

The chart in fig. 1.4 shows that, if we wish to study matter in atomic detail—for example to reveal the positions of the carbon atoms in a vitamin A molecule—we must use either X-rays, electrons or neutrons. The problem with electrons is that, so far, the only satisfactory way of eliminating the effects of the various aberrations of electromagnetic lenses is to make the aperture extremely small. The effect of aperture is discussed in more detail in the next chapter but we saw in Section 1.2 that the smallest details scatter radiation through the largest angles and hence if we cut down the aperture of the lens—or indeed of the pinhole or scanning hole—the information contained in the higher-angle scattering will not be accepted into the recombination system and so the finer details of the image will be missing. For electrons, therefore, the present-day limits on the perfection of available lenses prevent us from making use of their favourable wavelengths. As far as neutrons and X-rays are concerned no material or electromagnetic lenses are physically possible. Curved mirrors have been tried for X-rays but it is impossible to produce a magnification greater than a few hundred and even that is difficult. The magnification sought in order to study matter down to atomic dimensions (of the order of 10^{-10} m) is about 10^8 and hence this possibility may be rejected. What about the pinhole camera? We saw in fig. 3.4 that a pinhole camera for X-rays is feasible. Suppose we wish to reveal the atoms on the tungsten target what must we do? First of all we need to produce a magnification of the order of 10^8 and simple geometry shows that, even if we placed the pinhole say 0·01 m from the target, the film would need to be 1000 kilometres on the other side of the pinhole—and on that score alone the experiment is not very practical! But apart from that it is also clear that the pinhole would need to be comparable in dimensions with the detail to be imaged and this means a pinhole of atomic dimensions—which again presents problems! In other words pinhole camera systems can be ruled out for all the radiations in this region of the chart of fig. 1.4. Scanning also turns out to be impractical on the grounds that the spot of radiation which is scanned over the object must be of atomic dimensions and, at least for X-rays and neutrons, there are many other practical difficulties. For electrons—as we shall see later—scanning has certain very valuable practical advantages but these are not attainable anywhere near the dimensions which we are considering. The only direct possibility with electrons is the field emission idea mentioned in the last chapter but again this has very severe limitations.

The essential problem that remains, therefore, is that if we wish to 'see' detail on the atomic scale we must use X-rays, electrons or neutrons as our imaging radiation but there is no practical direct way of performing the recombination part of the process available to us at the present stage of our technological development. It seems therefore that we may be in a worse position than that of our audience in the third 'silly' experiment of Section 1.1 (page 2); we can irradiate the object under study and observe the scattered patch on the screen but we have no lens to place in the projector. Fortunately the practical position is not quite as bad as this and the equivalent of the patch of light on the screen scattered by the slide is rather more informative. Though the process of interpretation is indirect and may be very tedious, it is capable of yielding a good deal of information. The specialist subject described as X-ray diffraction or X-ray crystallography, which has been studied now for over 60 years, can be summarized as finding indirect ways of performing the recombination part of the process of imaging atoms by X-rays.

Why is it that the patterns are decipherable when the patch of light from the projector clearly is not? The primary reason is that the light used in an ordinary projector is both temporally and spatially incoherent whereas we saw in fig. 2.19 that if we use visible light which has some degree of spatial or temporal coherence, or both, the pattern ceases to be completely formless. Fortunately, it is possible to produce X-rays with reasonable degrees of both temporal and spatial coherence and hence scattering patterns which contain usable information may be produced. The second reason why a solution is possible is that the arrangement of atoms in almost all solid matter has at least some degree of regularity, and in a very high proportion the regularity is very marked. Thus the scattering patterns might be expected to be more like those of (*b*) or (*c*) in fig. 2.19, rather than those of (*d*) or (*e*). Fig. 4.1 shows two of the earliest X-ray diffraction photographs and it is obvious that they do continue a good deal of information.

It might be helpful at this stage to make a brief historical digression (though for a fuller account of the history of X-ray diffraction the reader is referred to Wykeham monograph No. 13, *Crystals and X-rays* by H. Lipson). The theme of the present book is the unity of the principles underlying imaging and diffraction techniques but it must be stressed that this unified view is only possible with hindsight. It has only begun to emerge over the last twenty years or so of the sixty odd years since the pattern of fig. 4.1 (*a*) was produced by Friedrich, Knipping and von Laue in Munich in 1912. Within a very few months of the discovery of the process a theory which accounted for the patterns had been worked out but it was in a form which did not readily lend itself to practical use. It fell to W. H. and W. L. Bragg a year or two later to develop the beautifully simple but revolutionary idea that led to the rapid development of the practical use of X-ray diffraction in revealing atomic detail. This was the concept enshrined in what is now called the Bragg Law. This was certainly the most important single break-through and led to an avalanche of new discoveries in metallurgy, chemistry, biochemistry, molecular biology and many other sciences. Yet, paradoxically, it placed such emphasis on the importance of regularity in the diffracting process that it may even have been one of the factors which delayed some of the later developments. In particular the realization that the interpretation of scattering from irregular objects is also possible and of the links between diffraction and microscopy seemed to take some time to emerge in their present powerful form. However, it was W. L. Bragg himself who first (in 1939) began to realize the power of regarding the techniques of X-ray diffraction as parallel with image-forming techniques in visual optics. In conformity with the theme of the book, the principles of the solution of X-ray diffraction problems will now be presented as a branch of physical optics, but of course this is only one of several approaches. It has a second advantage for our present purpose in that it can be presented visually through optical analogues and hence avoids the introduction of somewhat difficult mathematics.

Before discussing solutions let us look in a little more detail at the kinds of problems that may arise. The photographs of fig. 4.1 are both of the type called Laue photographs. They are produced by irradiating a crystal with a beam of X-rays and recording the scattered

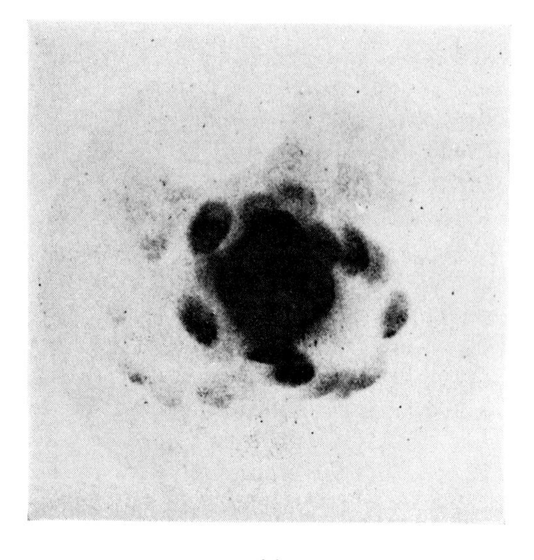

(*a*)

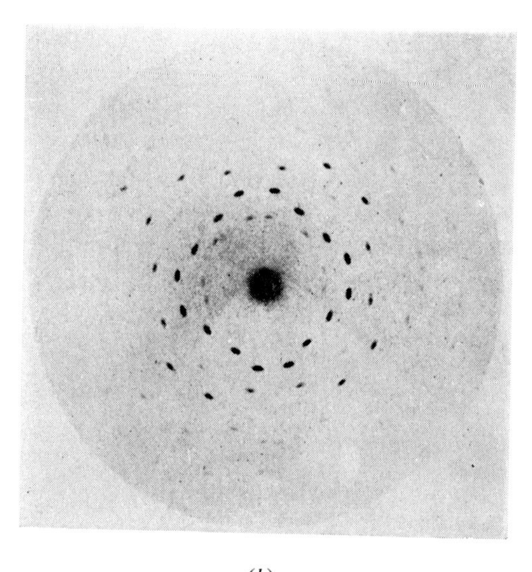

(*b*)

Fig. 4.1. (*a*) The first X-ray diffraction photograph; the crystal is copper sulphate. (*b*) An early X-ray diffraction photograph obtained by passing a fine beam of X-rays along an axis of symmetry of a crystal of zinc sulphide. From Friedrich, Knipping and Laue, *Sitzungsberichte der Königlich Bayerischen Akademie der Wissenschaften*, Munich, 1912.

radiation on film. The fact that the material was in the form of a crystal means that the object is extremely regular and von Laue and his colleagues thought of it as behaving like a three-dimensional diffraction grating. (Fig. 6.24, p. 163 is a beautiful electron micrograph of a virus crystal which illustrates the kind of regularity that occurs in crystals. In the case of zinc sulphide—the material for the photographs of fig. 4.1 (*b*)—the repeat unit, instead of being a complete virus unit with tens of thousands of atoms in it, consists of only eight atoms—four each of zinc and sulphur—and so is much too small to be revealed by electron microscopy, though the principle of regular repetition in three dimensions is the same).

The radiation in both photographs was ‘ white ’ radiation—in other words it consisted of a wide band of wavelengths just as visible white light consists of a wide range of wavelengths. In our terms it was temporally incoherent. For 4.1 (*a*) it is clear that the spatial coherence was not very great either—as evidenced by the large spots. (Fig. 2.19 (*m*)–(*r*) is an optical analogue which may help to clarify this point.) In fig. 4.1 (*b*) however the spatial coherence was greatly improved. The fact of temporal incoherence means that each spot has probably been produced by a different wavelength and, since we do not know what these wavelengths are, this greatly adds to the difficulty of interpretation. If fig. 2.19 (*m*) were in colour we should see that different parts of the pattern were in different colours and of course the colour would reveal information about the wavelengths contributing; in the X-ray case we have no such information. This kind of photograph, however, can be used for revealing information about the symmetry of the internal structure though, on the whole, it is less useful than photographs taken with temporally coherent (monochromatic) radiation.

Fig. 4.2 (*a*) shows a photograph for zinc sulphide taken by a so-called precession camera (fig. 4.2 (*b*)). It would be out of place to discuss the very complicated geometry of such cameras here. The complexity arises simply because we are dealing with details (atomic spacings) which are very similar in size to that of the wavelength of the X-rays used (0·15 nm) and hence the angles through which radiation corresponding to these details is scattered will be very large (refer back to fig. 1.3). Thus, ideally, we should record the information on a spherical film surrounding the crystal. The obvious practical problems of creating and using spherical films need no elaboration! The complex cameras used by crystallographers are various alternative solutions to the spherical film problem. The problem is comparable to the geographer's production of projections of the Earth's surface in an

(*a*)

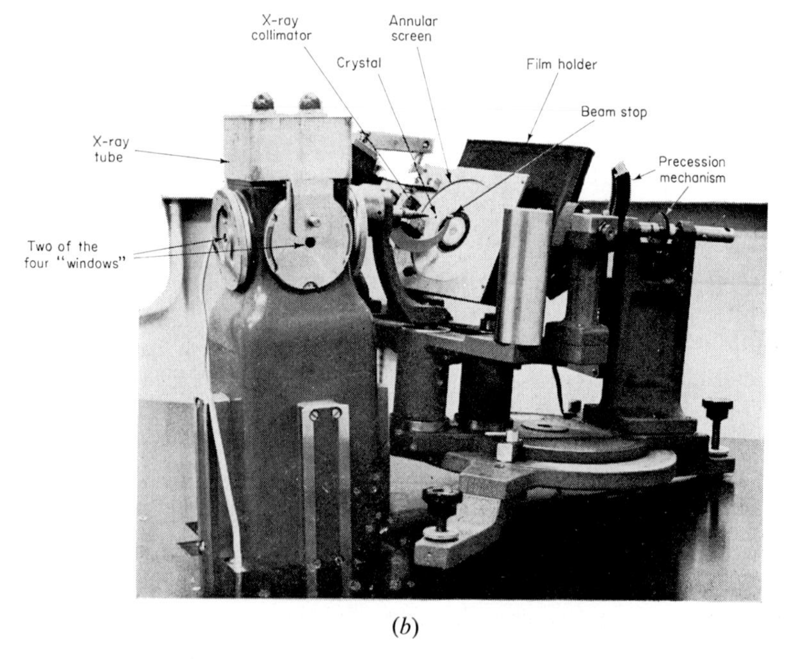

(*b*)

Fig. 4.2. (*a*) ' Precession ' photograph, an X-ray photograph in which the geometrical complexity lies in the camera and the pattern is relatively simple. The specimen is a crystal of zinc sulphide. (*b*) A precession camera.

atlas (Mercator's etc.) and permit the crystallographer to record the information he needs in a form that can be deciphered.

In the precession camera the crystal is made to precess about its axis and the flat film also undergoes a complicated motion which puts it in a position tangential to the ideal spherical film at the moment when any particular spot is being recorded.

When all the purely routine geometrical transformations have been sorted out, the photograph of 4.2 (*a*) can be shown to correspond very closely with the kind of pattern that would have been produced if

(*a*) a beam of radiation of wavelength 1/1000 of that used had

(*b*) been directed at the crystal and

(*c*) the pattern over quite a small range of forward angles had been recorded on a flat film (fig. 4.3).

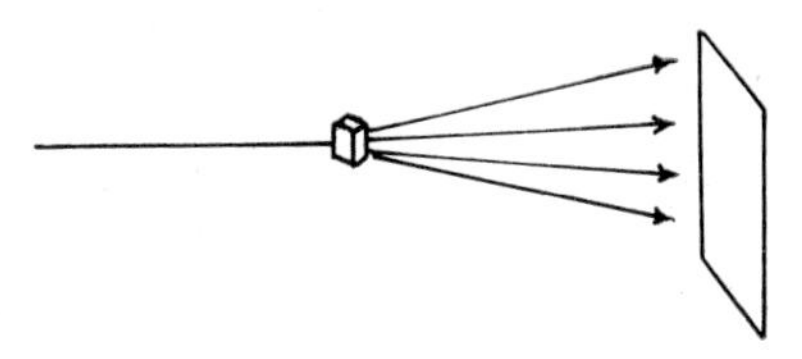

Fig. 4.3. Geometry for X-rays of wavelength (say) 10^{-13} m (10^{-3} Å) and atoms in zinc sulphide, or for light and holes a millimetre or so apart.

Except for the wavelength difference, the arrangement just described is very like those we used in producing optical diffraction patterns (e.g. fig. 2.15) and so we can illustrate X-ray diffraction patterns with an optical parallel or analogue.

Fig. 4.4 (*a*) shows a pattern made with visible light of wavelength approximately 7×10^{-7} m from a set of holes in a piece of card

(*a*)

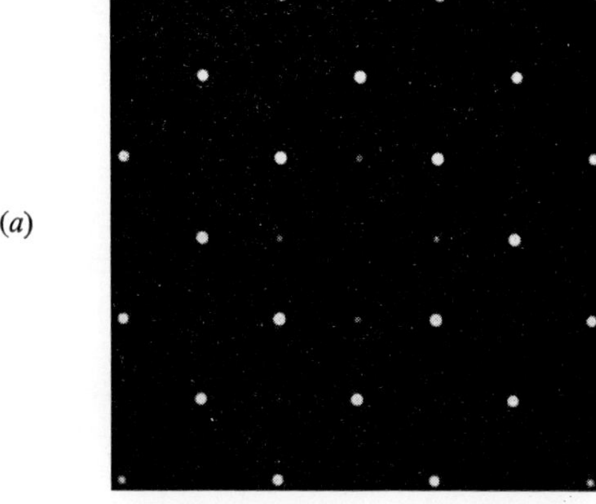

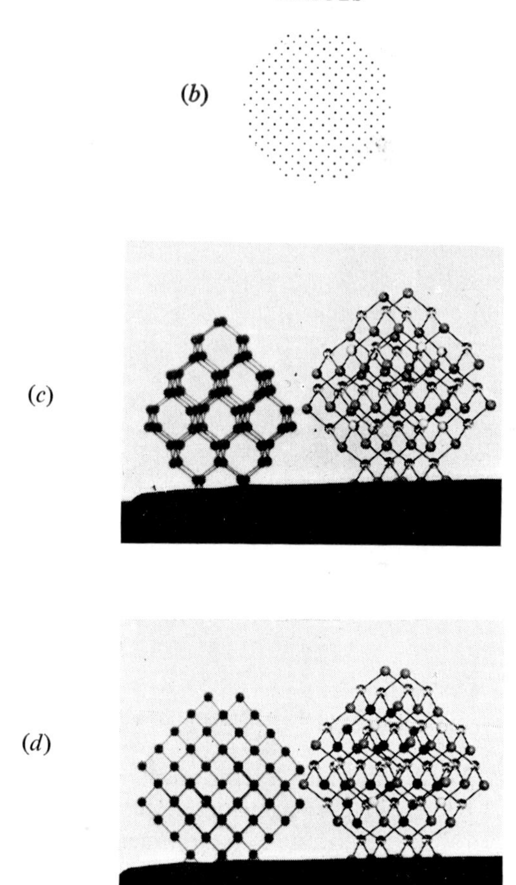

Fig. 4.4. (*a*) An optical diffraction pattern of a mask of holes representing the atomic positions in zinc sulphide with the central peak obliterated to match the X-ray photograph of fig. 4.2 (*a*). (*b*) The mask used for (*a*). (*c*) Model of zinc sulphide with shadow in a parallel beam of random direction. (*d*) Model of zinc sulphide with shadow in a parallel beam in the direction relevant to the photographs of 4.2 (*a*) and 4.4. (*a*).

(fig. 4.4 (*b*)). The holes correspond to the projections of the positions of the zinc and sulphur atoms in fig. 4.4 (*d*). The wavelength has been scaled up about 7×10^6 times from that specified in fig. 4.3. The actual atoms of zinc and sulphur are about $2{\cdot}35\times10^{-10}$ m apart and so to make a parallel we should have to scale this up to $2{\cdot}35\times10^{-10}\times7\times10^6=1{\cdot}6\times10^{-3}$; this is almost exactly the scale on which the mask of fig. 4.4 (*b*) was made.

Thus fig. 4.4 (*a*) should parallel fig. 4.2 (*a*) and indeed we can see that this is so. These pictures establish the feasibility of the optical analogue approach to the understanding of problems of interpreting X-ray diffraction patterns which will be elaborated in the next section. The optical diffraction patterns are produced on the apparatus illustrated in fig. 2.20 and, as in Section 2.4, we confine our attention to the detail in the central disc.

One further point may need clarification. I have referred several times to the ' projection ' of the positions of atoms. A study of the mathematical relationship—which involves three-dimensional Fourier transforms—shows that the photograph produced by the precession camera is related to the projection along the axis of the crystal perpendicular to the plane of the photograph. Figs. 4.4 (*c*) and (*d*) show a three-dimensional model of zinc sulphide placed in a parallel beam of light producing a projection as a shadow on the screen. For (*c*) the axis of projection is chosen randomly and in (*d*) the axis of projection is that relevant to the photographs of 4.2 (*a*) and 4.4 (*a*).

4.2. *Possible solutions using analogue methods*

Fig. 4.5 shows what happens if we start from one hole (which represents one atom) and add further ones to it to produce a complete hexagon (which can be thought of as representing a benzene ring—a regular hexagon of six carbon atoms of side about $1{\cdot}4 \times 10^{-10}$ m). It is important to realize that each separate hole gives the same pattern as that of fig. 4.5 (*b*) but the interference effects (which arise from the phase differences between the waves coming from the different holes) break up the pattern of the single hole into regions of smaller size. Suppose we now take this unit and repeat it regularly as a lattice. The result is as shown in fig. 4.6; the key point to notice is that the basic pattern is still further broken up into regions but the overall intensity of each region is related to the overall intensity in fig. 4.5 (*l*). This is a point of great fundamental importance since it shows us (i) that the position and spacing of the spots in the diffraction pattern of a regularly repeating object (i.e. in the X-ray diffraction pattern of a crystal) are determined by the way in which the basic unit repeats in the crystal; that is to the size and shape of the unit cell. But (ii) the arrangement of atoms *within* a unit cell (which is sometimes, but not always, a molecule or group of molecules of the material being studied) modifies the relative *intensity* of the spots of the pattern. Thus the two variables of the unit cell size plus shape and of molecular size plus shape may be determined separately. Fig. 4.6 also gives an indication of the way in which the transition from one unit to many affects the

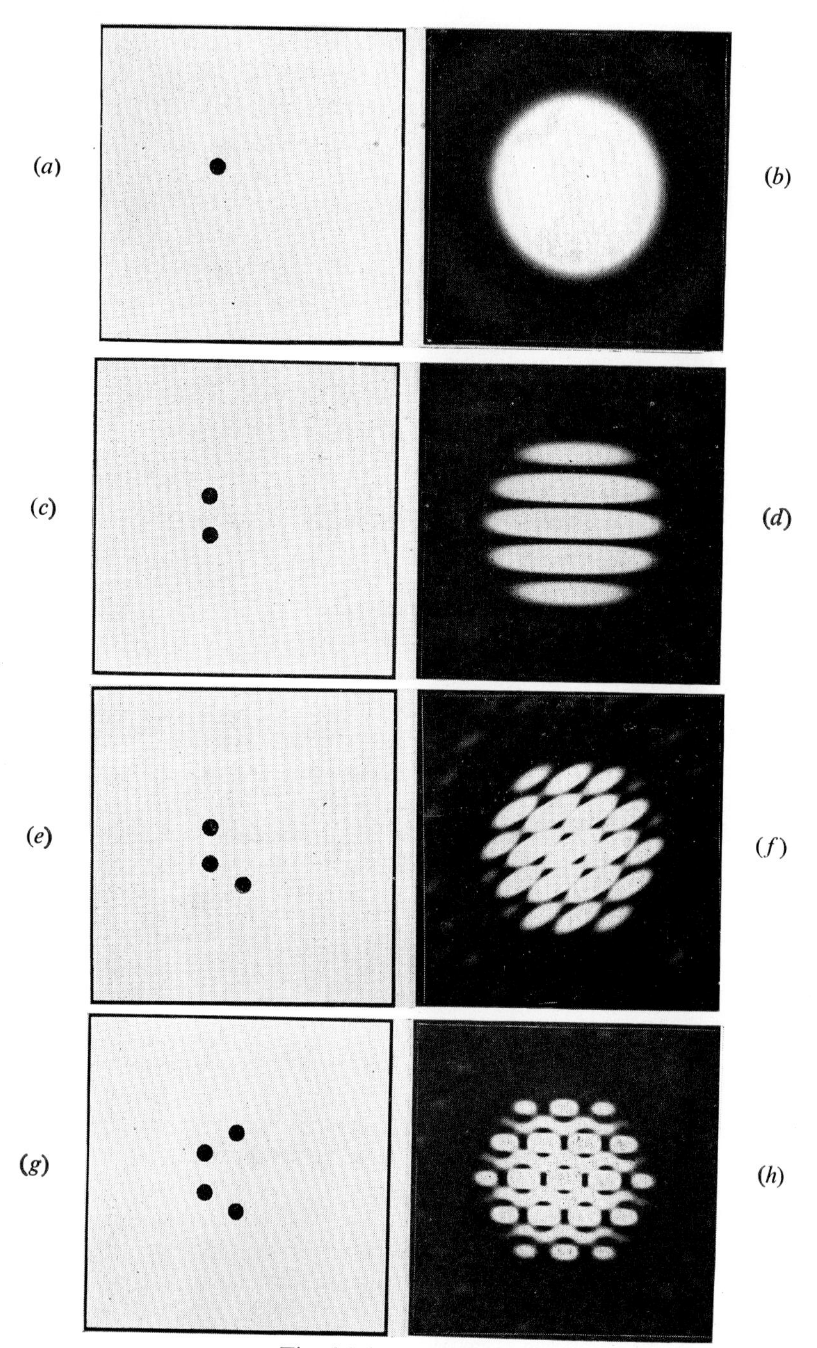

Fig. 4.5 (*cont. opposite*).

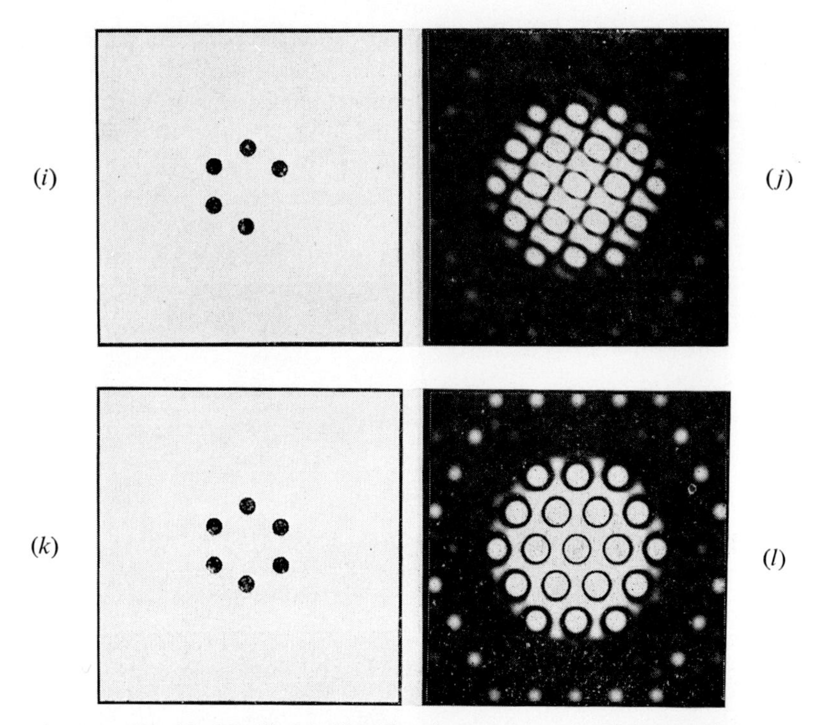

Fig. 4.5. (*b*), (*d*), (*f*), (*h*), (*j*), (*l*) Diffraction patterns of various combinations of holes forming parts of a simple hexagonal arrangement. (*a*), (*c*), (*e*), (*g*), (*i*), (*k*) The masks used. From *An Atlas of Optical Transforms*, by G. Harburn, C. A. Taylor and T. R. Welberry, by permission of G. Bell & Sons Ltd.

diffraction pattern. Figs. 4.6 (*e*) and (*f*) relate to a lattice with a fairly *large* unit cell so that the spots in the diffraction pattern are fairly close together. It is not too difficult, with half-closed eyes, to see the underlying diffraction pattern of one hexagon. In figs. 4.6 (*g*) and (*h*) the spacing in the lattice is smaller and much more like the kind of relative spacing that would occur in a crystal: it is much more difficult to see the diffraction pattern of the hexagon, and this ties up with experience with X-ray diffraction and crystals.

Four times in the last paragraph I have used the term ‘ unit cell ’ which is a piece of crystallographer’s jargon that perhaps needs explanation. Let us begin in two dimensions: fig. 4.7 (*a*) and (*c*) show two examples of repeating patterns that might occur in wallpaper. In each case we have drawn in the outline of a basic parallelogram by choosing identical features in the pattern. Notice that the same basic shape arises *whatever* point is chosen provided that the unit of pattern is repeated identically. We could reproduce the complete pattern by taking the contents of any one of the squares drawn in (*a*) or (*c*) and placing them down in rows and columns touching each other. Such a square would be called a unit cell. Notice that although its exact size and shape remain constant, it has no absolute position within the pattern—any point of the pattern will serve as the basis for

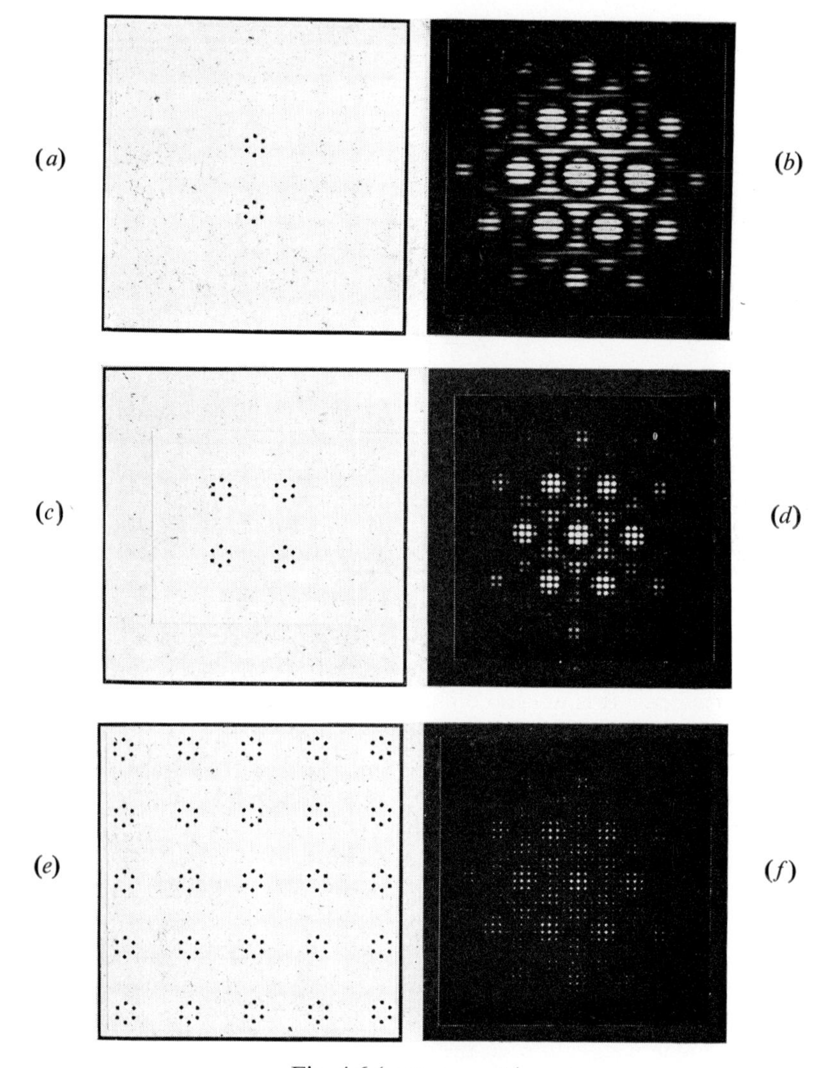

Fig. 4.6 (*cont. opposite*).

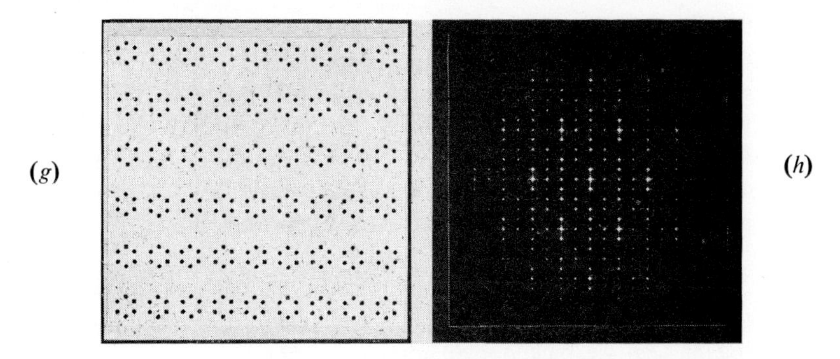

Fig. 4.6. Mask and diffraction pattern for: (*a*) and (*b*) two hexagons, (*c*) and (*d*) four hexagons, (*e*) and (*f*) Large numbers of hexagons on a square lattice, (*g*) and (*h*) Hexagons and a smaller lattice than in (*e*).

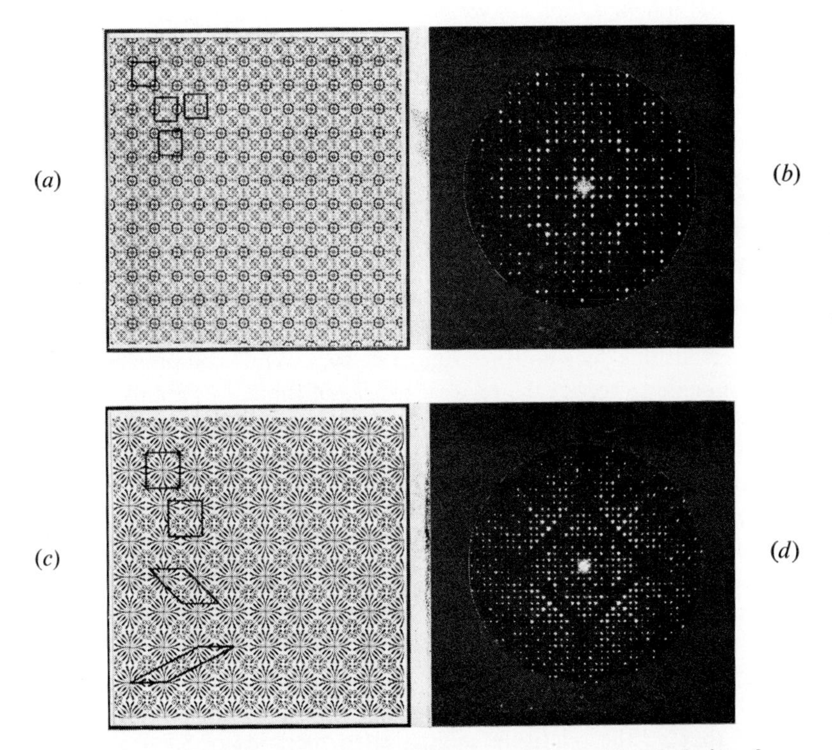

Fig. 4.7. (*a*) Wallpaper pattern showing alternative positional choices for a given unit cell size and shape. (*b*) Diffraction pattern of (*a*). (*c*) Wallpaper pattern showing alternative choices of shape of unit cell. (*d*) Diffraction pattern of (*c*).

the cell. Figs. 4.7 (*b*) and (*d*) show the diffraction patterns of the two wallpaper patterns and you can see immediately that the unit cell impresses its features on the diffraction pattern. You may notice too that the longer horizontal dimension of the unit cell in 4.7 (*c*) appears as a *shorter* horizontal dimension in 4.7 (*d*). This is just one aspect of the relationship between the real lattice and the lattice of the diffraction pattern that leads crystallographers to describe their relationship in terms of ' real ' and ' reciprocal ' space.

The extension to three dimensions is somewhat more complicated but follows exactly the same principles. In a three-dimensional repeating pattern we pick out an identical point in each repeat unit and, whatever point we pick, arrive at a three-dimensional lattice of points one parallelepiped of which is a unit cell. Notice however that there is not just *one* choice of unit cell. Even in two dimensions there are alternatives (fig. 4.7 (*c*) has one or two different units drawn in). Crystallographers adopt conventions in choosing unit cells but they are merely for convenience and to avoid confusion; there is nothing absolute about the choice and the problem need not worry us here.

Having established the basic relationships, we can now begin to see how X-ray diffraction patterns may be interpreted. The size and shape of the unit cell can always be determined directly from the position of the spots on the X-ray photograph—assuming that the wavelength and parameters such as the distance from the crystal to the film are known. But how may the relative intensities of the spots be related to the shape of the molecule or other unit? At first sight it may seem that it should be possible to calculate one from the other but this does not turn out to be so. Why? Simply because we are unable to record the relative phases of the X-ray beams and so part of the information is thrown away by the process of photography. As we saw in Section 2.6, there is no known way of recording the relative phases and so direct ' focusing ' of the scattering pattern by merely performing a computation is not possible.

You may recall that in Section 1.1 we pointed out that there are only two ways of focusing any image. One is to know all the parameters of the system (focal length of lens, distance from object and image to lens, etc.) and it can be shown that losing the phase information corresponds to losing this kind of information. The second possibility is to know what the object ought to look like and it is interesting to find that *all* the relatively sophisticated methods of solving X-ray diffraction problems, as well as the very straightforward ones, involve knowing something about the object, i.e. about the atomic arrangement in the structure being sought.

We shall consider very briefly some of the various techniques to see how this principle works out. The very simplest set of structures to solve are the so-called ' no-parameter ' structures. An example might be metallic tungsten, which consists of a repetition of just one atom at the corners of a rhombohedral cell with eight of its edges $2{\cdot}73 \times 10^{-10}$ m and four $3{\cdot}16 \times 10^{-10}$ m (though for their own curious reasons crystallographers often describe it in terms of one atom at the corner and one at the centre of a larger cubic unit cell (fig. 4.8) with sides $3{\cdot}16 \times 10^{-10}$ m). In this example the shape and size of the unit cell could be determined directly from the X-ray photograph, as we saw a few paragraphs ago. Then from our knowledge of the density of tungsten and its atomic mass, a very simple division of the atomic mass by the volume of the unit cell would tell us that there can only be one atom in each rhombohedral cell and so the structure is completely determined. In most cases however the content of the unit cell is more complex and the next simplest method is that of trial and error. By using knowledge of other similar compounds that have been solved already or by using chemical knowledge of what atoms are likely to be attached to each other, or by using knowledge of the sizes and shapes of atoms and how they can physically pack into a unit cell of a given size, or combinations of these, we can start to postulate a model. Chemical analysis and density measurements help us to deduce how many of each type of atom is in each cell and by using a knowledge of symmetry—which can often be determined from the X-ray patterns—we can devise a convincing model usually known as a ' trial structure '. This clearly comes in the category of ' knowing something about the object '. Then on the basis of this trial structure the relative intensities that it would give in the diffraction pattern can be calculated and compared with those actually obtained. If they agree then it seems

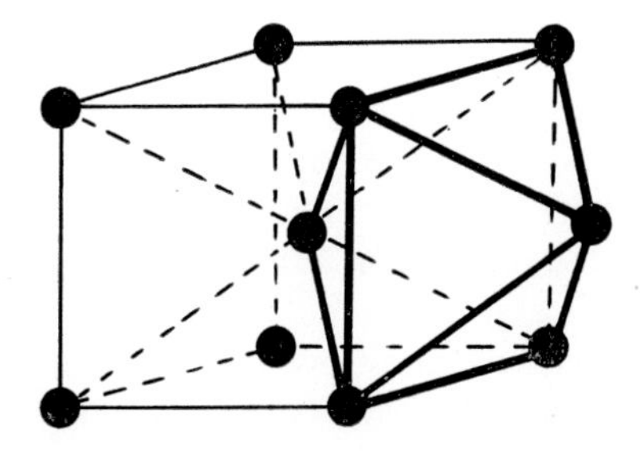

Fig. 4.8. Structure of metallic tungsten showing the relationship between the rhombohedral cell with one atom at each corner and the cubic cell with an atom at the corners and one at the centre.

likely that the proposal is in fact the solution; if they do not, modification has to be made until it does agree—hence the name ‘ trial and error ’.

How is the calculation actually done? Cast your mind back to the discussion in Chapter 2 centred round about fig. 2.10 in which we saw how a diffraction pattern could be built up from the fringes produced by pairs of points. The computer can be used to do just that: in effect each pair of points on the object produces a set of three-dimensional fringes which the computer samples at specific points and adds up the contributions of all the fringes. If three-dimensional sinusoidal fringes make your mind boggle, try to imagine one of those liquorice all-sorts which consists of alternate layers of black and white. Now imagine many more layers of much greater area. Finally imagine that instead of sharp black and white alternations the layers blend gradually into each other so that a plot along a line perpendicular to the planes would show a variation of intensity which was sinusoidal instead of square wave (like a set of battlements) as it would be for the original liquorice all-sort! Now you have imagined three-dimensional sinusoidal fringes.

A very powerful method that has been used to solve many of the more complex structures of biomolecular importance is known as the ‘ heavy atom ’ method. The technique is to prepare two forms of the crystal, one with the substance alone and a second in which each of the molecules has a ‘ heavy ’ atom attached to it. ‘ Heavy ’ in this sense means that it scatters X-rays very strongly—and since it is the electron clouds that do the scattering, this means one with a large number of electrons—i.e. a high atomic number. For example, a molecule that consisted of about 150 atoms of carbon, nitrogen or oxygen joined together (hydrogen atoms scatter so little that they can be ignored to begin with) each of which has six, seven or eight electrons might have a single mercury atom (with 80 electrons) attached to it at some point. X-ray patterns of the two compounds would then be compared, noting which of the spots gets brighter and which less bright when the heavy atom is present (these spots are called ‘ reflections ’ by X-ray crystallographers because of the Bragg Law notion of reflection from planes of atoms, but in our scheme of things this could be a confusing term).

This enables the relative phases of the spots to be determined under certain conditions and hence the structure can be ‘ focused ’ by computer.

A very remarkable and elegant aspect of our subject is bound up in that last ‘ throw-away ’ line; “ . . . and hence the structure can be 'focused' by computer." In Section 2.4 we said that the mathematical process needed to predict the diffraction pattern of a given object is known as Fourier transformation. Indeed the operation performed by the computer that was described only two paragraphs ago *is* that of Fourier transformation. The remarkable fact however is that *exactly the same process* is involved in predicting the object from its diffraction pattern. Mathematically speaking (and ignoring one or two provisos that ought to be made if we wished to be completely rigorous) the Fourier transform of the Fourier transform of an object is the object again.

Thus when we have the X-ray diffraction pattern we can divide it up into pairs of spots and each pair can be thought of as producing sinusoidal fringes which when added together will give us back an image of the object. The catch, of course, is that we must know the relative phases of the spots before we can perform this operation.

To indicate the way in which the heavy atom helps us to solve the phase problem consider fig. 4.9 in which all the pictures are optical analogues of X-ray diffraction patterns. Fig. 4.9 (*b*) shows the pattern for a single molecule of a compound called phthalocyanine and 4.9 (*d*) shows the pattern for a complete crystal. Fig. 4.9 (*f*) shows the pattern for one molecule with a 'heavy' atom (in this case rhodium) added and fig. 4.9 (*h*) shows the pattern for a complete crystal with the heavy atom. Only the centre regions are shown in (*d*) and (*h*) so that the changes in the intensity may clearly be seen.

How does this technique tie up with the idea that we must know something about the object? It corresponds in fact to the trick of focusing on the known feature—the chip in figs. 1.2 (*c*)–(*f*); in this case the heavy atom which is known to be present in each molecule is the chip and we focus on it and assume that the rest is then in focus.

The process of focusing by computer referred to above is usually called 'Fourier synthesis'—though it is probably more correct to call it Fourier transformation—and leads to electron density contour maps which are really the closest we can get to seeing the atoms in a crystal. Fig. 4.10 shows typical electron density maps that have been arrived at using the heavy atom technique.

Nowadays one of the most powerful methods of obtaining a solution is by so-called 'direct' methods. These give the illusion that one is merely taking the data and focusing an image by computation. Indeed there are now computerized systems which will accept the data and automatically calculate and even draw the 'image'. What has happened to the principle that one must know something about the object? How has the phase problem been solved?

Investigation shows that the mathematical relationships that are used to ' solve ' the phase problem are in fact based upon some very important facts about the structure which are implicit in the mathematical presentation. These are, for example, that the atoms are likely to be spherical, that the electron density is always a positive and real quantity—which in physical terms means that the electrons are behaving effectively as purely *mechanical* or geometrical scatterers and do not change the phase in any physical way. The remarkable

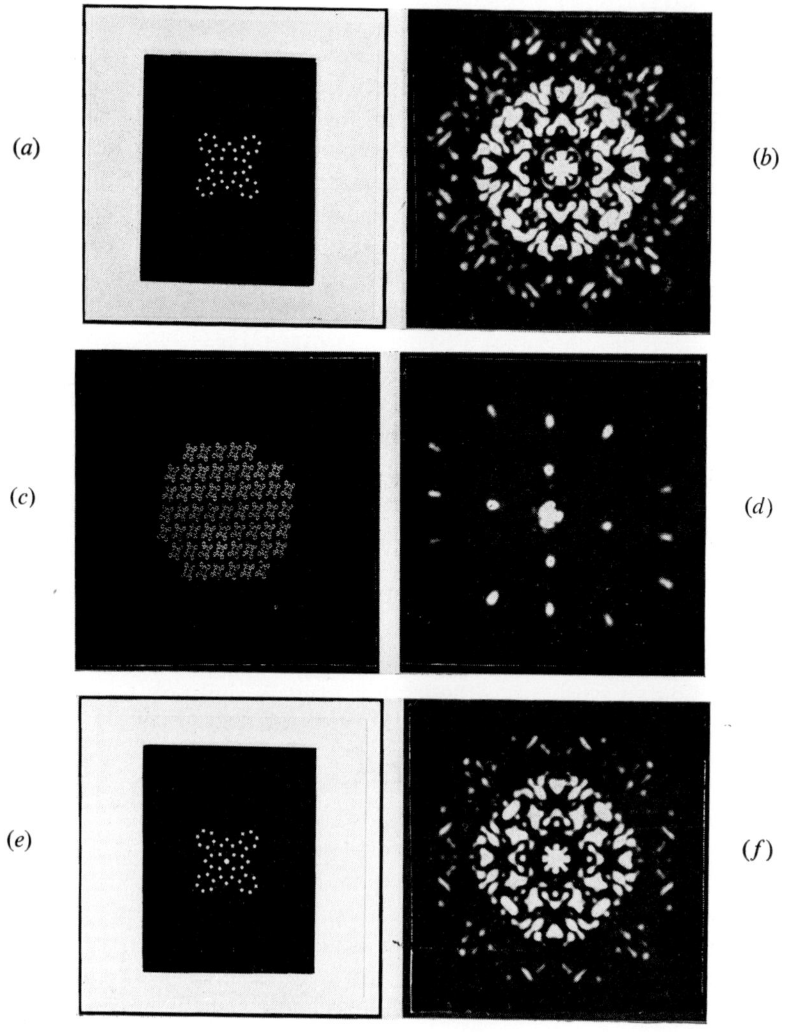

Fig. 4.9 (*cont. opposite*).

(g) 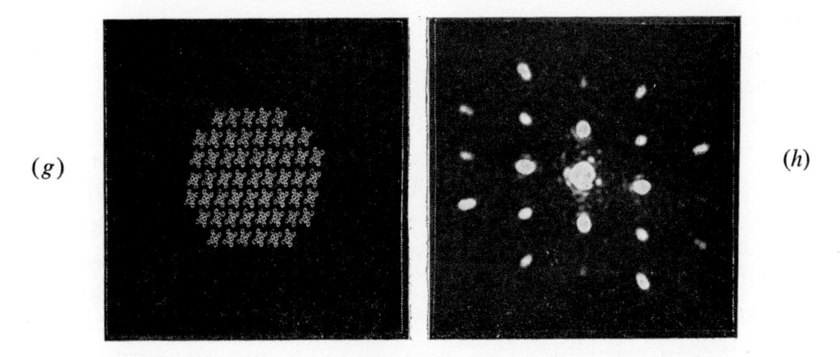 (h)

Fig. 4.9. Optical analogues illustrating the possible solution by the ' heavy atom ' method of the structure of phthalocyanine. (*a*), (*b*) Mask and diffraction pattern of one molecule with no metal atom. (*c*), (*d*) Mask and diffraction pattern of projection of crystal with no metal atom. (*e*), (*f*) Mask and diffraction pattern of one molecule with metal atom at centre. (*g*), (*h*) Mask and diffraction pattern of projection of crystal with metal atoms at all molecular centres. In (*d*) and (*h*) only the centre region is shown.

Fig. 4.10. Electron density map for nickel phthalocyanine. (Robertson and Woodward, *J. Chem. Soc.*, **219**, 1937).

thing is that these facts turn out to be powerful enough to give answers to a remarkably wide range of crystal-structure problems.

There are many other highly sophisticated techniques used in X-ray crystallography but perhaps those that we have selected will illustrate the principles sufficiently well for our purpose.

Is there really no way in which something looking more like an image of the atoms in a structure can be prepared? Yes there is—but it is only of use as an elegant curiosity since the trick can only be performed after the full details of the structure have all been determined. It does however provide a powerful confirmation of the general thesis of this book; that is, that all imaging processes can be divided into the two operations of scattering and recombination and for that reason an example is included here.

The technique was originally suggested by Sir Lawrence Bragg over thirty years ago and was perfected some ten years later in Manchester. The principle is simple; we use X-rays to perform the first stage of the imaging operation—the scattering stage—because of their small wavelength. Ideally we wish to convert the X-ray beams into beams of visible light of the same relative intensities and phases. If this could be done we could then use an ordinary lens to focus the image. The problem lies of course in the conversion of the X-ray beams into light beams.

Once the intensities of the X-ray spots have been determined, the structure solved, and the relative phases calculated from the known structure, a plate containing holes to represent each X-ray spot is prepared. Each hole carries a piece of mica and the imaging process is carried out in polarized light. If a piece of mica of suitable thickness is rotated in its own plane between crossed polaroids or nicol prisms the light transmitted rises and falls in intensity and passes through a zero of intensity every 90°. (See any text book on physical optics or crystal optics.) By rotating the mica in each hole, the intensity can thus be varied to match the relative X-ray intensity associated with the spot; it turns out that if the mica is rotated clockwise from a position of zero intensity the phase of the light passing through is 180° different from that if the mica is rotated anti-clockwise. Fortunately, if the structure is centro-symmetrical (that is for any point with coordinates x, y, z with respect to some origin there is always a point $-x$, $-y$, $-z$) the only phases that occur are 0° and 180° and hence this technique provides all that is necessary. In fact further refinements of the technique led to the possibility of producing any required phase but it would be too long a diversion from our theme to discuss the details here. (See for example Lipson, *Optical Transforms.*) Fig. 4.11

is an example of an image produced in this way and shows a simple molecule and parts of its neighbours in a crystal of hexamethylbenzene in which the inner hexagon is the benzene ring of six carbon atoms and the outer hexagon is the six carbons of the methyl groups. The hexagons do not appear to be regular because we are viewing them obliquely. One could legitimately describe this photograph as a photo-micrograph with a magnification of about 10^8; but we must remind ourselves that it cannot be obtained directly and is merely a way of presenting the result when the structure has been solved by one or other of the techniques of X-ray crystallography.

4.3. *Holography*

In Section 1.1 the term ' hologram ' was used to describe the patch of light on the screen formed by a projector without its lens. We said

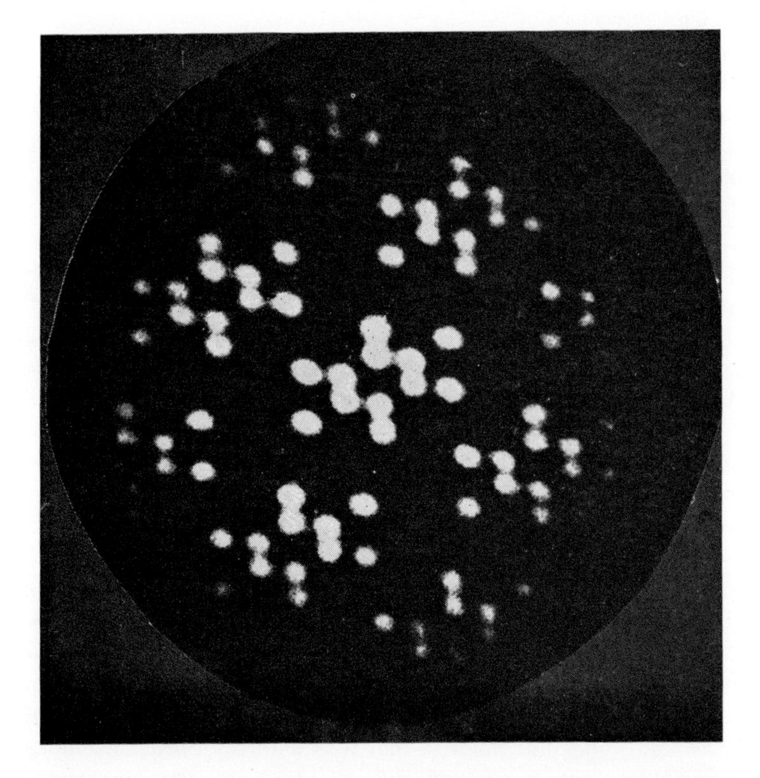

Fig. 4.11. Image produced by the two-stage process of scattering X-rays and then recombining light beams which can only be done when the structure has already been determined. The structure here is hexamethylbenzene.

there that the reason is simply that *each* point on the screen is receiving information about *every* point on the object. The word is derived from the Greek *holos* meaning ' a whole ' and of course reflects the notion that the whole of the information about the object is contained in it. This particular kind of hologram is not, however, very useful mainly because of the incoherence of the light. We saw in fig. 2.19 that the scattering patterns of objects become more detailed and contain more information when the degree of coherence of the radiation increases and it seems feasible therefore to expect that it would be easier to extract information from a hologram if it were made with coherent light. Various partially successful attempts were made before 1960 to recombine the information contained in holograms without using a lens but the real breakthrough came with the development of the gas-phase laser; in one step there was an almost unbelievable increase both in temporal and spatial coherence and incidentally in total light intensity.

We shall first of all examine the simplest form of experimental arrangement used for producing holograms and for reconstructing images from them and then consider two simple alternative approaches to understanding how the technique works. Fig. 4.12 (*a*) shows the experimental arrangement for producing the hologram.

The object O is illuminated by a temporally and spatially coherent beam from the laser L_1. The primary beam from a gas phase laser is usually only a millimetre or two in diameter and so it is normal to expand the beam by means of a lens system. It is essential that the lenses used should be of fairly high quality so that there is no disturbance of the spatial coherence. Light is scattered from the object and falls directly on to a photographic plate P. A portion of the original beam from L_1 is directed by the mirror M so that it produces a uniform patch of light over the whole of the plate P. Without going into detail at the moment, it is clear that a complex interference pattern will be produced on P as a result of the superposition of all the various scattered waves and of the so-called reference beam from M.

In order to reconstruct the image the developed photographic plate is illuminated by a beam from a laser, L_2 in fig. 4.12 (*b*), in such a way that the beam falls on the plate in the same direction as that of the reference beam from the mirror in the first stage. An observer looking into the plate from E will then see an image of the object at I and the striking and important point is that it is a three-dimensional image. In other words if the eye is moved from side to side the same parallax effects occur as would with a real object. The effect is particularly pronounced if the object consists of two or more separate items some

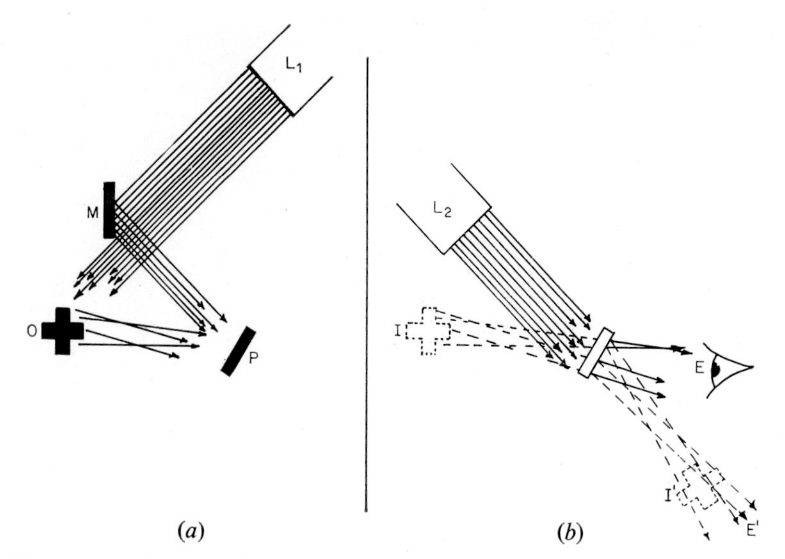

Fig. 4.12. (*a*) Schematic arrangement for producing a hologram. (*b*) Schematic arrangement for reconstructing an image from a hologram.

of which are nearer to P than others. Fig. 4.13 shows two views of such a reconstructed image from different directions, and the parallax effect is most pronounced.

How can the process be explained? Let us first consider a rather elementary approach. In fig. 4.12 (*a*) we have a set of waves scattered from O which have very precise phase relationships with each other by the time they reach the plane of the photographic plate at P. The reference beam is simply a uniform wave front also arriving at P. The two sets of waves will be superimposed on each other and an interference pattern will be produced which may be recorded on the plate. In the reconstruction stage, 4.12 (*b*), this interference pattern recorded on the plate is introduced into the replica of the reference beam produced by L_2. The effect as far as the eye E is concerned is that the light from L_2 has been modified to resemble precisely the superposed interfering waves at P in fig. 4.12 (*a*). You will recall that we met the idea that all optical images are really interference patterns and the eye and brain accept these waves to form an image in exactly the same way that they would accept the interfering waves from a *real* object placed at I. The effect to the observer is therefore precisely as though a full three-dimensional object were actually at I.

This explanation may or may not convince you: it is reasonable as far as it goes but it leaves a great many questions unanswered and of

Fig. 4.13. Two views of a reconstructed image from a commercially produced hologram illustrating the three-dimensional character of the image.

course a really full explanation demands quite sophisticated mathematics. The following treatment takes us a little further, however, and may be found useful.

* Let us suppose that the object O is replaced by a single scattering point. The waves scattered from it (remember that it is illuminated by temporally and spatially coherent light) will simply be a regular train of spherical wave fronts diverging from the point. The intersection of these spheres with the plane of the photographic plate will be a set of concentric circles. Suppose that we ' freeze ' the whole system in time and that we consider only the spherical surfaces at intervals of 1 wavelength where the phase has some particular value. Then the intersection with the plane of the plate will be a set of concentric circles which get closer together as we move away from the centre (see fig. 4.14). Thus if we now superpose the uniform reference beam the resulting interference pattern on the plate will look very like a zone plate (Section 3.3) except that the zones will not have sharply defined edges. Now, on transferring to the reconstruction stage (fig. 4.12 (*b*)) we have a zone plate in a parallel beam of light and hence a point image will be produced (a real one to the right of the plate and a virtual one to the left). The same argument may be followed through for every point on O and we can then see that the hologram is really an immensely complex superposition of zone plates each of which will produce an image of its own corresponding point on the object.

Two questions immediately jump to mind. What about the multiple images produced by a zone plate? In the last chapter we said that a new kind of zone-plate lens could now be produced by holographic methods which had only one focus on either side. Indeed the explanation we have just given is at the heart of the new technique. It is the substitution of a sinusoidal variation for the square wave variation in blackness of the rings that, in fact, removes the higher orders. The effect is very much the

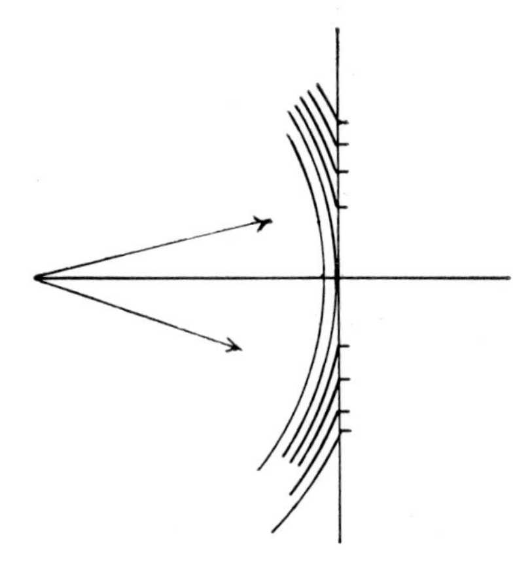

Fig. 4.14. Diagram showing interaction of spherical waves with plane of photographic plate.

same as the fact that a diffraction grating in which the variation of density along a line perpendicular to the grating lines is sinusoidal, instead of a square wave, gives only one order of diffraction on either side of the centre instead of many.

This last point is an important one and can most easily be approached by thinking of the reversibility of light waves—or the two-way application of the mathematics of Fourier transformation.

Consider first of all the diffraction pattern of a double slit: it is simply a cosinusoidal variation in amplitude. (We say *co*sinusoidal here because there is a *maximum* at the centre as for a cosine curve). Fig. 4.15 (*a*) illustrates this. Now suppose we add a third slit on the axis, midway between the other two and suppose, although still very narrow, it transmits twice the amplitude of the other two. Think back to our consideration of the geometry of the double-slit on page 33 and you will realize that this third slit is on the axis and therefore contributes in phase with the beam from point O in fig. 2.13 everywhere on the screen. The resultant pattern is thus as shown in fig. 4.15 (*b*)—the cosinusoidal curve is lifted so that it is everywhere positive and the former negative regions just become zero.

Now suppose we make a grating whose transparency distribution is like that of fig. 4.15 (*b*); the principle of reversibility of light waves or the Fourier transform relationship tells us that *its* diffraction pattern will be the equivalent of the three slits. That is, it will be a central order with one order on either side each of half the amplitude of the centre one.

Now to return to the zone plate view of the hologram. What about the real image? It turns out in fact that there *is* a real image from a

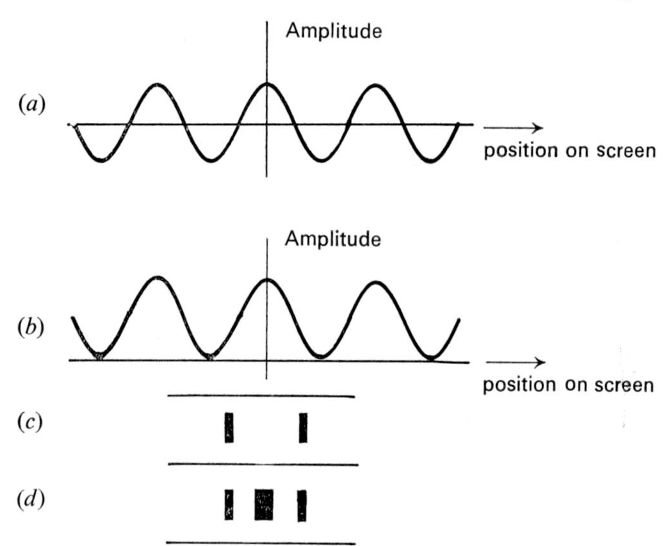

Fig. 4.15. (*a*) Cosine amplitude distribution from double slit. (*b*) Cosine amplitude distribution with constant term added resulting from triple slit. (*c*) Double slit producing (*a*). (*d*) Triple slit producing (*b*).

hologram and this can be picked up and photographed just as easily as the virtual one. In fig. 4.12 (*b*) it would be at I′ and could be photographed by a camera placed at E′. It is not quite so easy to observe the real image because the waves are diverging from it and a camera of limited aperture would have to be very carefully aligned in order to see it and the field of view tends to be more limited.

Holography has some very interesting applications in many different areas of physics and a section of the last chapter is devoted to two of them (6.13).

5. Perturbations of the image

> " Yes, I have a pair of eyes," replied Sam, " and that's just it. If they was a pair o' patent double million magnifyin' gas microscopes of hextra power, p'raps I might be able to see through a flight o' stairs and a deal door; but bein' only eyes, you see, my wision's limited."
>
> Charles Dickens
> *The Pickwick Papers*, ch. 34

5.1. *Aperture and wavelength*

We have already hinted once or twice at the possible effects if the object is so small compared with the wavelength of the radiation that its diffraction pattern involves large angles which take much of the radiation outside the aperture of the instrument; but we must now look in a little more detail at the consequences. First let us consider a simple circular aperture and the problem of imaging an object which consists of a fairly regular arrangement of holes of different sizes. We shall use the instrument described in Section 2.5 (fig. 2.20 (*a*)) with lens L_3 in position so that an image of the object placed at P is produced at R. If we start with the instrument operating at full aperture, most of the diffracted waves will enter the system and contribute to the image and hence the image is almost identical with the object (fig. 5.1 (*a*)). Its diffraction pattern (recorded at Q) is shown in fig. 5.1 (*b*). Now if we introduce an aperture at Q which cuts out some of the diffracted beams and only allows the portion shown in fig. 5.1 (*c*) to enter the system, the image at R becomes like fig. 5.1 (*d*). Notice immediately

Opposite, above:

Fig. 5.1. (*a*) Object (mask) of holes in opaque card. (*b*) Optical diffraction pattern or transform of (*a*). (*c*) Portion of (*b*) allowed to go forward for recombination. (*d*) Recombined image from (*c*).

Opposite, below:

Fig. 5.2. (*a*) Smaller portion of fig. 5.1 (*b*) selected for recombination. (*b*) Recombined image from (*a*). (*c*) Still smaller portion of fig. 5.1 (*b*) selected for recombination. (*d*) Recombined image from (*c*). Figs. 5.1 and 5.2 from *An Atlas of Optical Transforms*, by G. Harburn, C. A. Taylor and T. R. Welberry, by permission of G. Bell & Sons Ltd.

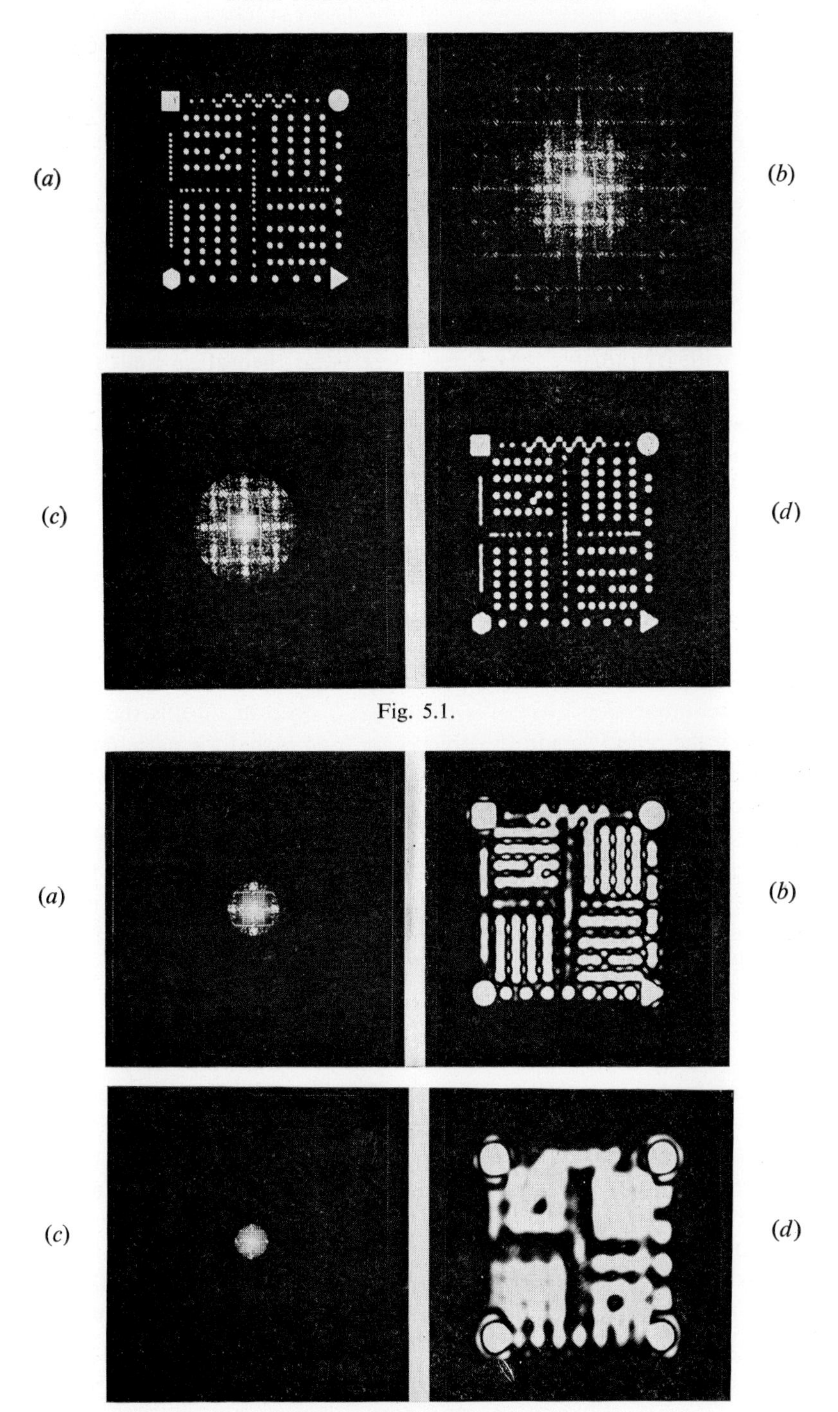

Fig. 5.1.

Fig. 5.2.

that it is the finest detail that is affected; some of the very closely spaced holes have become blurred into a line and in the case of the shapes at the corners of the object the sharp angles have become rounded. This process continues as the aperture is still further reduced (fig. 5.2).

In practice these limitations occur in all systems whether using visible radiation or not. This is one of the reasons why astronomical telescopes, both visual and radio, are made as large as possible—the other main one being that of course a large aperture gathers more light waves and hence is likely to produce an image that is easier to see. Since the stars we observe are so far away, each one bathes the whole Earth with its radiated waves and there is no theoretical limit to the improvement obtained by increasing telescope size—a telescope with an aperture equal to the Earth's diameter could in theory pick up all the radiation falling on the Earth and hence give the best image! Obviously there are other practical considerations that make limitations long before this size is reached. In the microscope, however, the situation is different. The objective lens of the microscope is so close to the object that it is possible to design a lens which will take in almost all the cone of scattered radiation from the object and hence the theoretical limit of resolution is much more closely approached with microscopes than with telescopes. One of the reasons why the so-called oil-immersion objectives (in which a drop of oil fills the space between the objective and the object) are used is that the oil minimizes the effect of the boundaries between the cover slide and the lens and so increases the cone angle of light actually entering the system, thereby increasing the resolution.

I have talked in terms of changing the size of the aperture—but a moment's thought and a further study of fig. 1.3 shows that it is the size of the aperture *relative to the wavelength* that matters and that doubling the wavelength for the same aperture is the same as halving the aperture with the same wavelength. For visible light we only have a wavelength range of between about 4×10^{-10} m and $7{\cdot}5 \times 10^{-10}$ m so we can barely double the wavelength even if we go from deep purple to deep red. In other regions of the electromagnetic spectrum, however, big changes of wavelength can be made. If we are designing a radar set, therefore, a smaller wavelength will not only suit us better from the point of view of allowing us to encode more detailed information about the object, but will also allow us to use smaller reflectors. In radio telescopes, however, the choice of wavelength is made by the stellar radiating sources and we have to design our dishes of a suitable size to give the required resolution with the given wavelength (which might be of the order of 20 m).

It will help us not only now but also in later work if we give some attention to the more detailed aspects of the behaviour of apertures. Consider, for example, a telescope used by an observer to look at a single star. The star is so far away that the spherical waves radiated by it will have a radius of curvature that is quite a reasonable approximation to infinity! The telescope is thus effectively being irradiated by a parallel beam of light. The lens, of course, has the effect of bringing this parallel beam to a point focus in its back focal plane (assuming that the lens is perfect) and so we have the ideal conditions for observing Fraunhofer diffraction (as specified in Section 2.5). Thus if we placed a small circular aperture over the lens, we should see the aperture's Fraunhofer diffraction pattern when we look through the telescope. Now suppose we enlarge the aperture. The diffraction pattern of a circular hole is the famous Airy disc (fig. 5.3 (*a*)) and as the hole gets larger the pattern stays the same in form but is reduced in scale. (Some authors retain the name 'Airy Disc' for the central peak only: I find it more convenient to use it to refer to the whole pattern of central disc and concentric rings.) The radius of the central disc is, in fact, inversely proportioned to the radius of the hole (fig. 5.3 (*b*)). What happens when the aperture becomes sufficiently large to coincide with the rim of the lens? Clearly we still have an Airy disc, albeit a small one, when we look at the star through the telescope. Indeed this is the smallest disc we can obtain with the given aperture of the telescope. In other words we can *never* see a true image of the star but will always see the Airy disc of the aperture. This is the origin of the comment in Section 3.3 that, when considering an image as an interference pattern, the notion that the phases other than at image

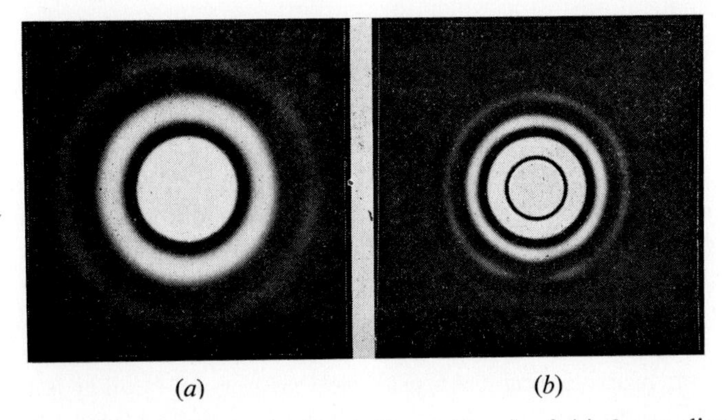

(*a*) (*b*)

Fig. 5.3. Diffraction patterns (Airy disc patterns) of (*a*) 2 mm diameter circular hole, (*b*) 4 mm diameter circular hole.

points are such that they result in zero amplitude is not true at points close to the image point.

Now suppose there were several stars in the field of view. Precisely the same argument would apply to each and we should have an Airy disc at each of the points corresponding to a star. They would all be the same *size* though their relative brightness would depend on the brightness of the stars. (Fig. 5.4 (*a*) is a picture of a star field and the Airy discs can be seen very clearly.)

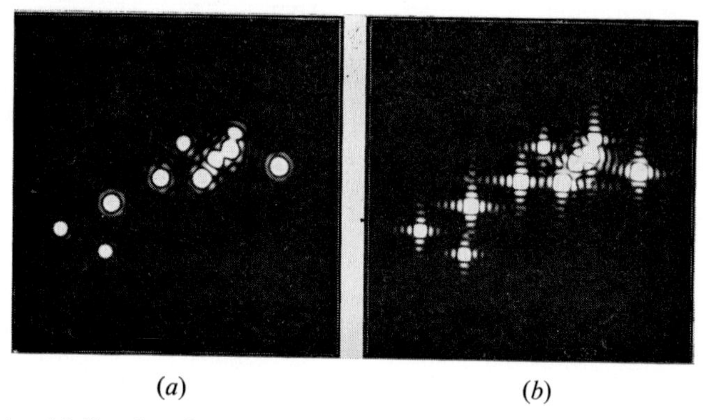

(*a*) (*b*)

Fig. 5.4. (*a*) Simulated star-field with Airy pattern at each point as it would be imaged in a telescope of small aperture. (*b*) The same star-field imaged with a rectangular aperture over the telescope aperture.

The process of replacing each point image of a star by an Airy disc of identical size and form is an example of a very important concept in Fourier transform theory called ' convolution '. Convolution is a mathematical operation which for our purposes can be thought of as a process of dealing a particular item to a series of positions. An example from everyday life might be the regular arrangement of eggs in a moulded tray ready to be packed in a crate. We could describe the arrangement of the eggs as the convolution of one egg with a set of mathematical points arranged on a square lattice. We could describe the star field as shown in fig. 5.4 (*a*) as the convolution of two items; one is a picture of the star field with mathematical point images and the other is a single Airy disc. The fascinating point that emerges is that this phenomenon of convolution always occurs in diffraction patterns when two objects are ' multiplied '. That last clause may take a bit of explaining. An example that may be familiar is that of ' crossed ' diffraction gratings. A diffraction grating placed in a laser beam gives a single row of regularly spaced orders in a direction perpendicular to the lines. A second identical grating placed on its own in the beam with its lines perpendicular to those of the first will also give a regular row of orders, this time in a direction at right angles to those of the first. If now both gratings are placed in the beam so that the beam goes first through one and then through the other the 'crossed' gratings are effectively multiplied; the diffraction pattern now consists of a set of orders at the points of a square lattice—which can be described as the convolution of the set of orders from one grating with the set of orders from the other (see fig. 5.5). That this operation is multiplication and *not* addition is also illustrated in fig. 5.5. When the two gratings are placed side by side the result is the single addition of the two separate transforms translated, as always, to a common centre. The fact that the addition depends on *phase* as well as amplitude is demonstrated by the fringes in the central region where the two patterns overlap; these fringes are characteristic of the lateral separation of the two gratings.

In the case of the telescope, the lens aperture is bathed in the scattering pattern of the star field. All the parts of this scattering pattern outside the aperture are lost completely and this can be regarded as 'multiplication by nought'. Thus if we think of the aperture as a function of magnitude one within the circle and of magnitude nought outside it, you should be able to see that we have indeed multiplied the scattering pattern of the star field by the aperture.

To understand the idea of convolution completely takes a long time but I hope the basic idea will begin to make sense soon, as it can simplify enormously our approach to some of the later problems.

We have already seen quite a few examples in earlier parts of the book. In Section 4.2 we talked about the separation of the variables of lattice dimensions and unit-cell contents. In our new terminology the crystal can be thought of as the convolution of a lattice of mathematical points with the contents of one unit cell. Now we have already met the reciprocal nature of Fourier transforms and it should not be too surprising now to be told that this time the diffraction patterns are multiplied. Fig. 4.6 in fact shows this very clearly; fig. 4.6 (*f*) can be seen to be the *product* of two items, one a perfect lattice with every point of the same intensity (which is the diffraction pattern of the lattice of mathematical points), and the other, the diffraction pattern of the unit cell contents (fig. 4.5 (*l*)).

Now we should be in a position to answer the question, "What would happen if we were to make a telescope with a *rectangular* aperture instead of a circular one?" Clearly we should now be multiplying by a rectangular function and so our star field would be

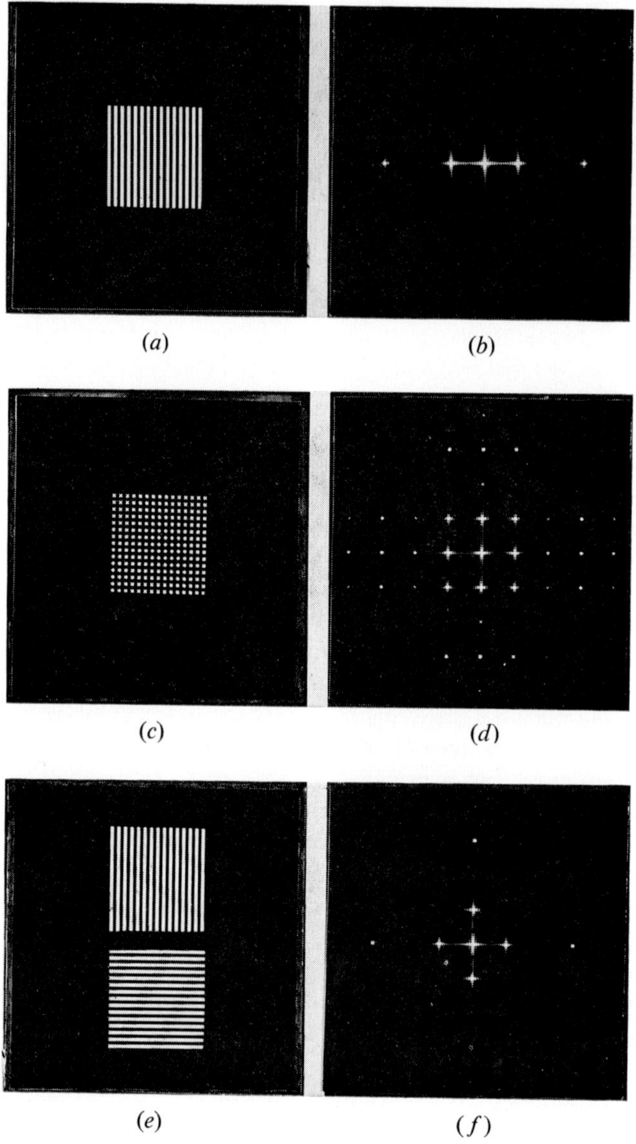

Fig. 5.5. (*a*) A diffraction grating of rectangular slits. (*b*) Optical transform of (*a*). (*c*) Cross grating produced by ‘ multiplying ’ the grating of (*a*) by a second identical grating turned through 90° and placed on top of (*a*): light is only transmitted where *both* are clear. (*d*) Optical transform of (*c*): it is the convolution of (*b*) with itself turned through 90°. (*e*) The grating of (*a*) ‘ added ’ to a second identical one turned through 90° by placing them side by side. (*f*) The optical transforms of (*e*): the transforms also are added.

the convolution of the point-field with the diffraction pattern of the rectangle. Fig. 5.4 (*b*) shows the star field in the telescope with the rectangular aperture. The diffraction pattern of a rectangle now appears at each star point.

One final remark needs to be made. Telescopes are, of course, used to observe many objects other than star fields, but the convolution principle remains valid. Every point of an extended object in fact becomes an Airy disc if the aperture is circular and the result may be quite complicated. It is considered in more detail later in the next section.

5.2. *False detail and possible misrepresentation near the resolution limit*

Almost since the microscope was invented there have been controversies about the fineness of detail that could be observed. Usually the physicists have been on one side of the fence, claiming—along the lines that we have already discussed—that it is impossible to image detail whose significant dimensions are much less than the wavelength of the radiation used; on the other side of the fence have been the practical microscopists who have said " But look down this microscope —I can actually *see* detail that is smaller than the wavelength ". What is the solution to the paradox? As so often is the case *both* claims are true. The microscopist *can* see detail that is smaller than the wavelength but it may very well be totally false and may not correspond to anything that exists on the object. Thus in the end the physicist is also right if he changes his claim to indicate that it is impossible to make a *reliable* image of detail whose significant dimensions are much less than the wavelength of radiation used.

How can this finer detail arise? Fig. 5.6 (*a*) shows a pattern of four points arranged on a square. Fig. 5.6 (*b*) shows this pattern as it would be imaged by a system with a limited circular aperture. We obtain this pattern by using the convolution process described in the last section and placing an Airy disc pattern at each of the four points. Now suppose that the four points move closer together (fig. 5.6 (*c*)): the Airy discs will also move closer together and a position can be found when the first ring of each of the four Airy patterns all intersect at the centre of the square (fig. 5.6 (*d*)). The result will be a bright spot at the centre and we shall ' see ' a spot that is not there. Because the rings of the Airy pattern have a thickness which is less than *half* the diameter of the central disc, this additional spot will be smaller than the discs and may well appear smaller than the normal limit of resolution.

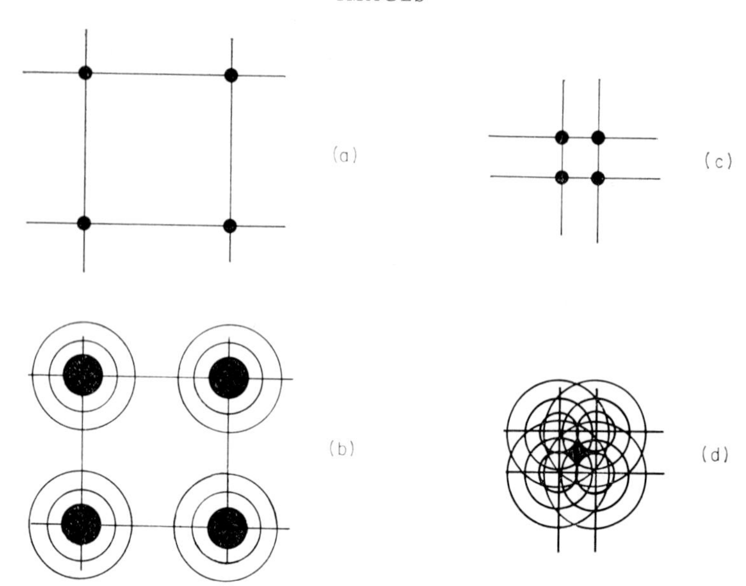

Fig. 5.6. (*a*) Four points arranged on square. (*b*) Airy disc patterns at each of the four points: the lines represent zeros and the central discs are shown black. (*c*) Four points much closer together. (*d*) Airy disc patterns of the same size as those in (*b*) placed at the points of (*c*). The central diamond shaped patch shown black is the superposition of the first bright rings from all four patterns: note how much smaller it is than the central discs in (*b*).

Fig. 5.7 shows some examples of the kinds of spurious effects that can occur. Fig. 5.7 (*a*) is the object and (*b*) is its diffraction pattern. When an attempt is made to form an image with a small aperture (*c*) which allows only a limited amount of the diffraction pattern to enter the system the result is (*d*). The condition referred to in the last paragraph and illustrated in fig. 5.6 has been deliberately created for the 5×5 array of holes at the centre; the additional ' holes ' which are much smaller than the Airy discs can clearly be seen. In (*e*) the aperture is still further restricted and the result at (*f*) is spurious in yet a different way. We now have a 3×3 array at the centre in place of the real 5×5 array!

Fig. 5.7 alone is an awful warning to microscopists not to believe all that they see, especially near the limit of resolution for their particular system. That, in turn, means that microscopists—or indeed any users of image-forming systems—should always *know* what the resolution limit of their system is.

A very beautiful practical illustration of the effects was produced some years ago by Mr. A. W. Agar—it is shown in fig. 5.8. On the

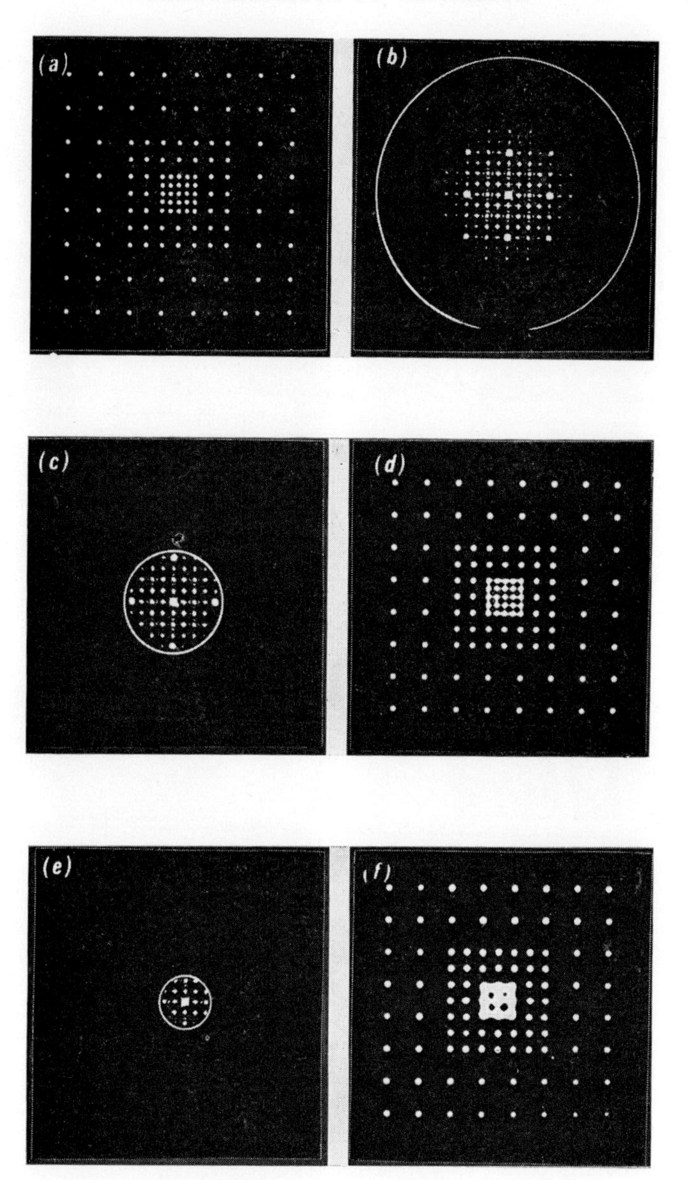

Fig. 5.7. (*a*) Regular object. (*b*) Optical transform of (*a*). (*c*) Restricted pattern of (*b*). (*d*) recombined (*c*): note particularly the 'extra' holes which have appeared in the central square as illustrated by fig. 5.6. (*e*) Further restrictions of (*b*). (*f*) recombined (*e*): note that the centre square now appears to be 3×3 instead of 5×5. From *Optical Transforms*, by C. A. Taylor and H. Lipson, by permission of G. Bell & Sons Ltd.

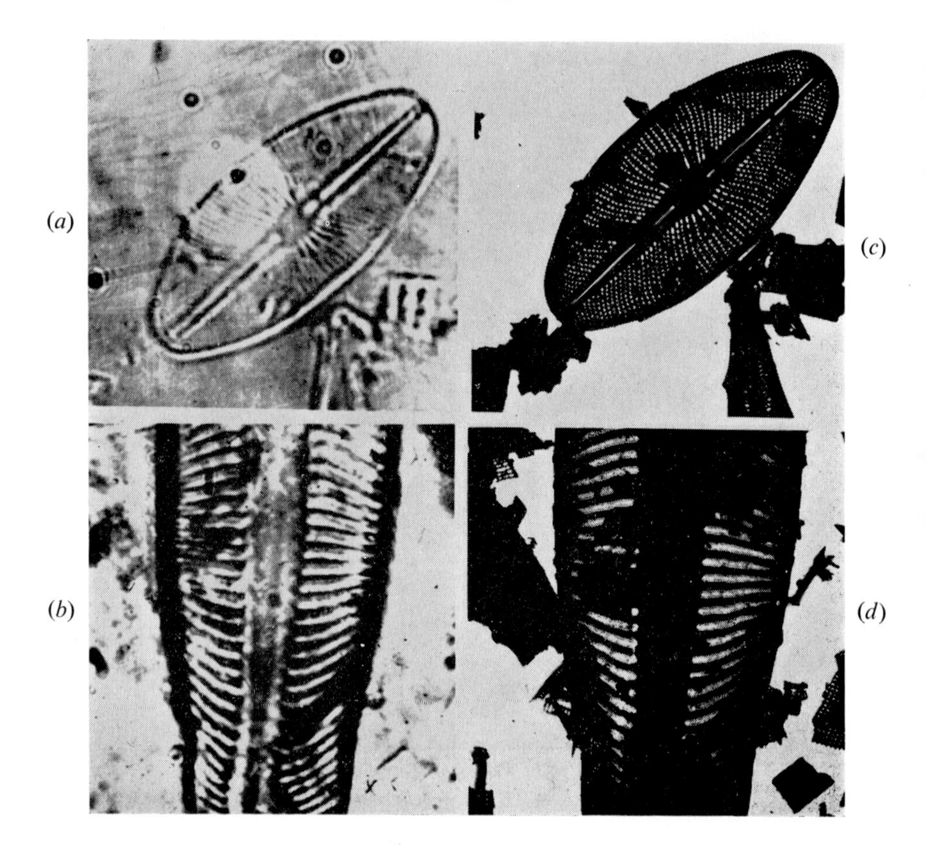

Fig. 5.8. (*a*), (*b*) Optical micrographs of two different diatoms showing detail close to the resolution limit. (*c*), (*d*) Electron micrographs of the identical specimens at the same magnification in the electron microscope. By permission of Mr. A. W. Agar, Bishop's Stortford, Herts.

left are two optical micrographs of some diatoms in which the fine detail is beyond the resolution limit. On the right, by extremely elegant manipulation, *electron* micrographs—for which the detail is well within the resolution limit—have been made of the identical areas of the diatoms so that the true detail can be clearly seen.

The effect does not *only* hold for regular objects, and fig. 5.9 shows the effect of aperture reduction on the image of Mickey Mouse! Again it is the fine detail and sharp edges which suffer. In modern optical jargon it is common to say that the ' higher spatial frequencies ' are excluded. This is of course a mixture of ideas and relates to the model put forward in Section 2.3 about the image being built up from sets of fringes; the object with the highest spatial frequencies—that

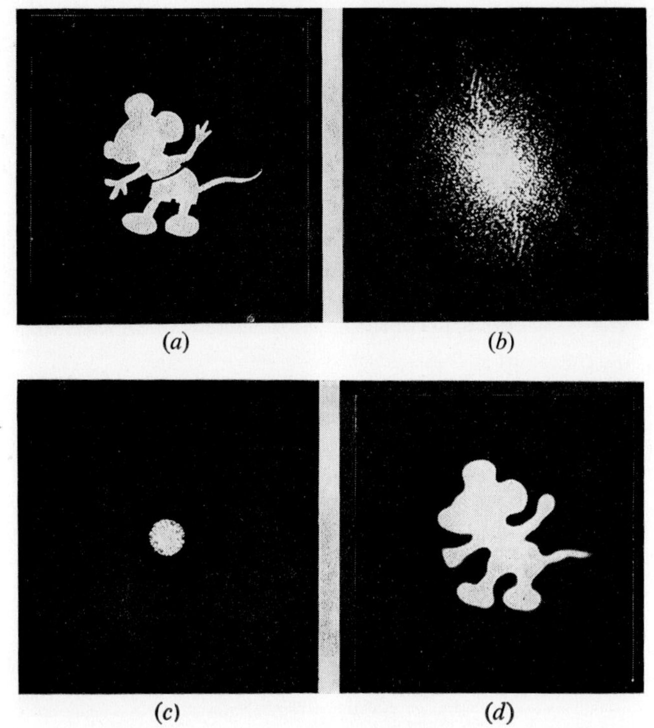

Fig. 5.9. (*a*) Outline of somewhat irregular object. (*b*) Optical transform of (*a*). (*c*) Restricted portion of (*b*). (*d*) Recombination from (*c*). From *An Atlas of Optical Transforms*, by G. Harburn, C. A. Taylor and T. R. Welberry, by permission of G. Bell & Sons Ltd.

is with details very close together—produces fringes with the widest spacings and these are likely to be the first to go when the aperture is reduced.

We can still use the notion of convolution, however, and in fig. 5.10 we have deliberately chosen objects which, though continuous rather than being made up of isolated points, nevertheless display in recognizable form the features of the Airy discs with which each element of the object is replaced. In fig. 5.8 the optical micrograph of the diatoms show fringe effects which arise from the subsidiary rings of the Airy discs.

5.3. *Abbe's theory of microscopic vision*

The problems that we have discussed in the last two sections worried the microscopists of the 19th century very considerably and it was not

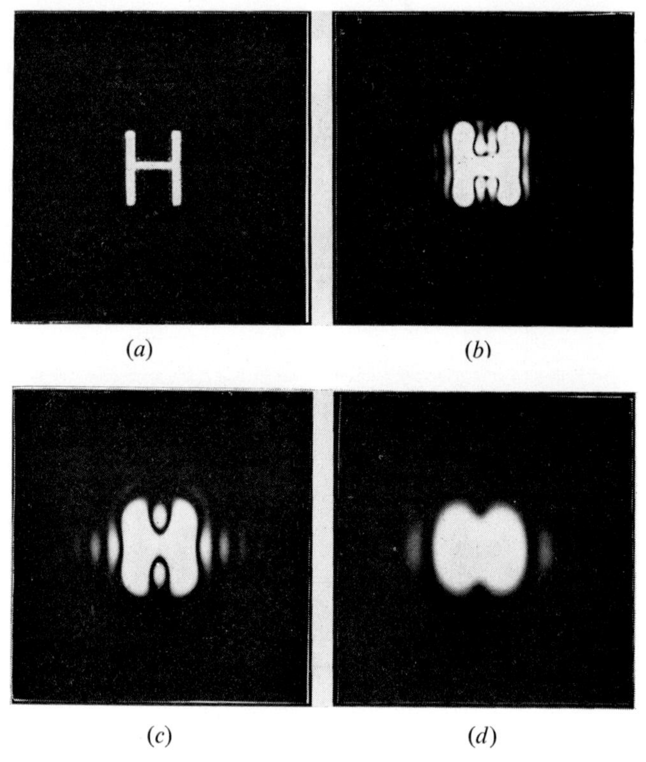

Fig. 5.10. (*a*), (*b*), (*c*) and (*d*) Four recombinations of successively reduced portions of the optical transform of a transparent letter H.

until the German physicist Abbe produced his famous theory that the behaviour near the resolution limit began to be understood. The convolution approach described in the last sections has really superseded Abbe's ideas but, partly for historical reasons, and partly because of the elegance and simplicity of the idea—which remains correct though somewhat limited in application—we shall devote a little space to it.

Abbe began by considering just how the image of a regular diffraction grating of parallel rulings might be produced in a microscope. Fig. 5.11 shows the kind of ray diagram with which his arguments have been illustrated. We show only the objective of the microscope—the eyepiece has little effect on the resolution problem. The real image of the grating G appears at G′. But any set of parallel rays from the grating slits will come to a focus in the back focal plane of the objective before going on to form the image. The back focal plane will thus contain the orders of diffraction of the grating. The diagram shows very clearly how the scattering pattern in the back focal plane then goes on to become the image. By means of a series of extremely elegant experiments in which Abbe actually cut out various orders of diffraction in the back focal plane of his microscope objective, he was able to study the effects on the image—rather as we did in a different way for figs. 5.1 and 5.2.

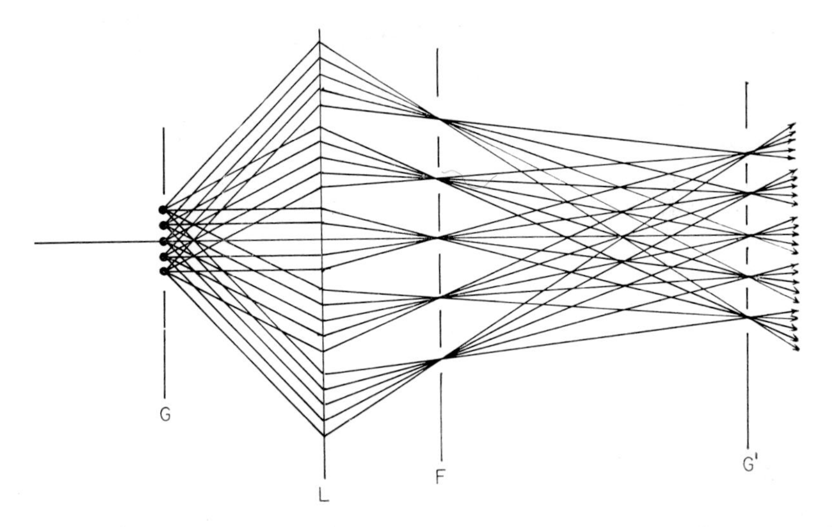

Fig. 5.11. Ray diagram illustrating Abbe's theory of the microscope. G is the regular grating, L the microscope objective, F the orders of diffraction in the back focal plane of L and G′ is the image of G.

On the basis of his experimental evidence he formulated the principle that the final image in a microscope will not be an exact replica of the object but will be the replica of an object which would have as its diffraction pattern that portion of the diffraction pattern that is allowed to enter the image-forming system. Thus if we take fig. 5.2 (*a*) as an example: the image obtained by letting this into an image-forming system will be of an object which has 5.2 (*a*) as its total diffraction pattern. And this is precisely what 5.2 (*b*) is. For a regular object, the periodicity in the image will disappear when only the zero order of the diffraction pattern is allowed to contribute to the image.

Abbe recognized that his theory applied to irregular as well as to regular objects and really laid all the foundations for modern image-processing techniques—which we shall discuss in more detail in the last chapter. As I said at the beginning of this section, his theory is largely superseded by the Fourier transform–convolution approach but it is very good practice to think through some of the examples we have given in terms of his theory. Indeed one of the delights of this kind of optics is that very often there are several different approaches that can lead to a given result. You begin to have a much more confident feeling about a subject if you can arrive at the same result by more than one method.

5.4. *Image 'formants'—the fingerprint of the apparatus*

In the study of sound there is a very important principle that each time a sound wave passes through any kind of system, the system will impose its own fingerprint or formant characteristic on the sound. For example, our voices are created first by the vibration of the vocal cords. The rather harsh buzzing sound then passes through the throat–nose–mouth system with all its cavities and at least two distinct formant characteristics are imposed on it. There is a formant characteristic for each vowel sound, and this is the same whoever makes the sound, and there is a formant characteristic that distinguishes one person from another. A further example would be the curious things that happen to the sounds created in a concert hall when they are transmitted by radio. The differences between the sound reproduced by a super hi-fi system, a moderately good portable radio and a miniature transistor radio are considerable. In each case the receiver has imposed its own formant characteristics. Some of you will have tried to build amplifiers and will know how difficult it is to produce uniform amplification at all frequencies, and it is the departures from uniformity of amplification that we call the formant characteristic.

The parallel with what happens in image-forming systems is remarkably close. The idea of diffraction patterns being made up of summations of fringes has been mentioned several times and we have also talked about the reciprocal relationship between the spacing of two points of an object and the spacing of the corresponding fringes in the diffraction pattern. In terms of Abbe's ideas, the finer the detail on the object, the further from the axis would be the corresponding order of diffraction, and it is clearly these higher orders that are likely to be cut out in an imaging system. Thus it is likely to be the high spatial frequency element of the pattern that will be the first to go—as indeed we saw in the case of Mickey Mouse in fig. 5.9. The elimination of spatial frequencies above a certain level would be a simple kind of formant that would occur wherever an image is formed by a system with a limited aperture. But the effect is much more general and we find that all kinds of variations in the frequency content can occur on the way through a system and each element will impress its own formant characteristic on the image. Thus if we are concerned with the transmission of a television picture, it must now be clear that electronic distortions or modifications of the signal between transmitter and receiver, or in the receiver itself, will change the picture. Indeed, we can generalize Abbe's theory and say that the final picture will be of an object that would have produced the signal that is finally fed to the picture-forming circuits rather than of the object itself. Another example of a way in which a formant might arise is in any system in which scanning is involved. If the spot, instead of being small, becomes larger, or becomes elongated, or indeed any shape other than a mathematical point, then the image will be the convolution of the true image with this shape. This, in Fourier transform terms, corresponds to multiplying the scattering pattern by the diffraction pattern of the deformed spot and so the result is just the same as for a microscope or telescope with an odd-shaped aperture.

It may seem that this section is becoming somewhat philosophical, but the principles being expressed are of great importance. Perhaps they will become clearer in the last chapter, when we shall look at a great many examples of how our ideas work out in practice. It is particularly important to try to understand the principles in order to study various systems for image correction or image enhancement, such as are used for example in processing television images returned from spacecraft.

So far we have talked about image-forming systems in which we assumed the lenses to be free of all aberrations. In fact, lenses are rarely free of aberrations. They may for example suffer from spherical aberration—which means that a spherical wave incident on the lens does not merely change its curvature but ceases to be spherical after passing through the lens. It cannot therefore come to a point focus and any image produced by it will be distorted. There are other aberrations that can arise which we will not consider in any detail. It will suffice to say that all can be regarded as formants of the system and, in some cases, we can correct the defects by compensating for the unwanted formants.

5.5. *Methods of measuring the performance of imaging systems*

In modern optics it has become customary to specify the overall performance of an imaging system in one of two ways. It becomes necessary to specify it because, if we are producing an image of something we have never seen before (e.g. a new strain of bacteria) we cannot compare the image with the object because we do not know what it *should* look like. We therefore need some kind of objective criteria.

One of the standard methods is closely related to the formant idea described in the last section. We consider each spatial frequency in turn and find out how well the system reproduces it. The result is a function called the transfer function; it describes how each frequency is transferred through the system and what happens to it when it reaches the other side. In a system free from all aberrations (a so-called diffraction-limited system) the transfer function would simply be of unit magnitude up to the maximum spatial frequency admitted to the system and zero outside this. For systems with aberrations the function becomes more complicated and may involve phase as well as amplitude changes.

The second standard method is to use what is called the point spread function. This is simply the image on one side of the system resulting from a mathematical point source on the other side. Again for a diffraction-limited system the result is quite simple: the point spread function is simply an Airy disc. For systems with aberrations it might be quite a complicated shape and may vary from point to point of the object. In Section 6.9 the method of studying electron microscopy essentially involves the experimental determination of this point function and fig. 6.28 shows the complex form that it can take when aberrations are present.

5.6. *Depth of focus problems*

There is one image defect that is worth picking out from the rest for special consideration before we proceed to practical applications, and that is a manifestation of the focusing problem which has already cropped up once or twice. We have tended to think of two-dimensional images but often they are produced from three-dimensional objects. Simple geometry shows that it is unlikely that we can devise a lens that will produce a perfectly focused two-dimensional image of a three-dimensional object, and this supposition is amply borne out in practice.

Fig. 5.12 is a series of photographs which illustrate the phenomenon. We start with a photograph of a young lady and successively move closer until in 5.12 (*e*) we have just about reached the limit with an ordinary camera; the origin of the effect is discussed in section 6.3. It is very obvious that only one plane in the texture of the fine wool scarf is being reproduced satisfactorily. We then transfer to the microscope and in 5.12 (*f*) and (*h*) increasing magnification shows finer and finer detail and the problem of depth of focus remains with us.

Is there a solution? There is, but it involves completely different imaging system which depends on scanning rather than on a simultaneous transformation using lenses. We have illustrated the effect here by transferring to the scanning electron microscope and in figs. 5.12 (*g*), (*i*), (*j*) and (*k*) greater and greater detail is shown without any depth of focus problems. The flakes that can be seen in the pictures at higher magnification are odd specks of face powder on the scarf!

How does the scanning system solve the depth of focus problem? The point is simply that the scanning beam is a fine pencil of electrons (or of light in the flying-spot microscope) which traverses the specimen and at any one moment the detector is receiving information only

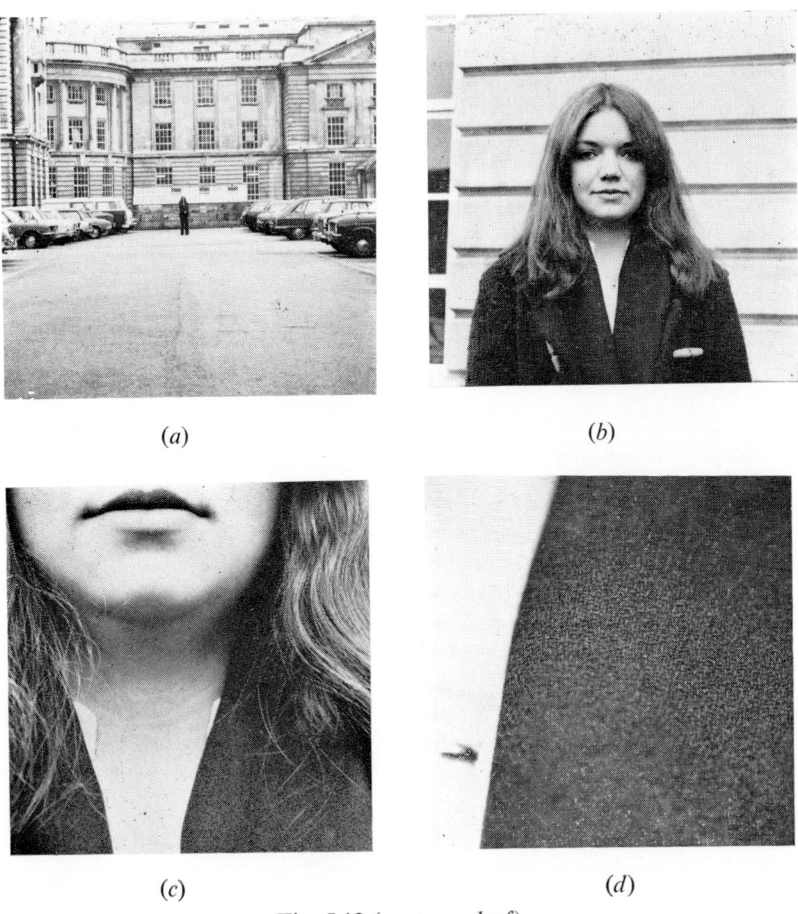

(*a*) (*b*)

(*c*) (*d*)

Fig. 5.12 (*cont. overleaf*).

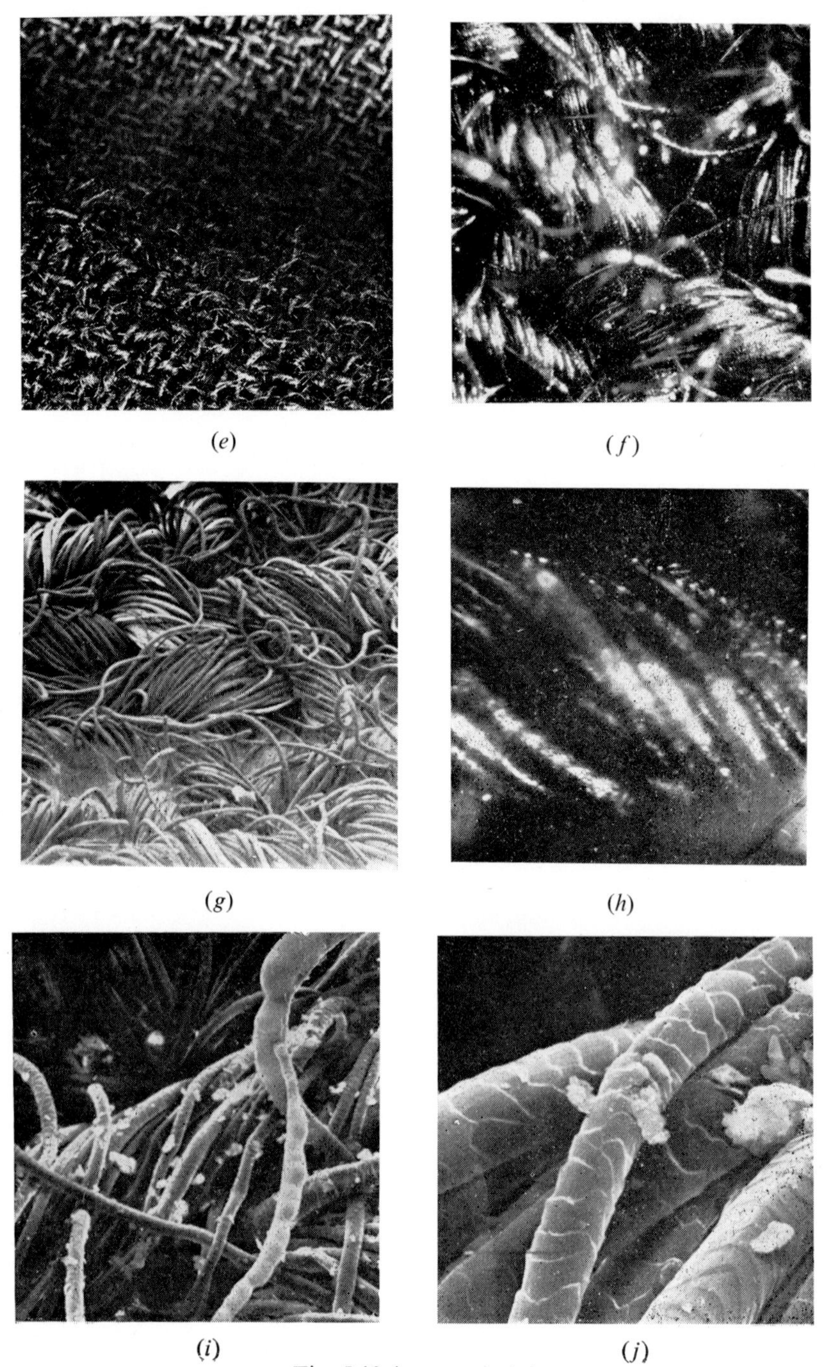

(*e*) (*f*)

(*g*) (*h*)

(*i*) (*j*)

Fig, 5,12 (*cont. opposite*),

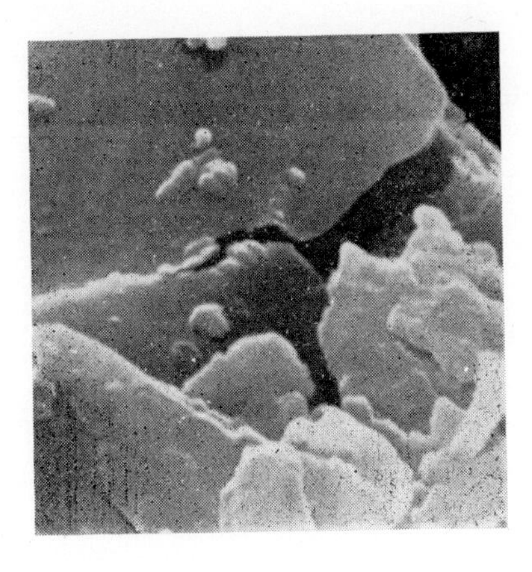

(*k*)

Fig. 5.12. (*a*)–(*e*) Photographs with an ordinary camera at closer and closer range. (*f*) Optical micrograph approx. ×25. (*g*) Scanning electron micrograph approx. ×25. (*h*) Optical micrograph approx. ×100. (*i*) Scanning electron micrograph approx. ×100. (*j*) Scanning electron micrograph approx. ×600. (*k*) Scanning electron micrograph approx. ×6000. The pale flakes in (*i*), (*j*) and (*k*) are traces of face powder. The regular scales are well-known features of wool fibres.

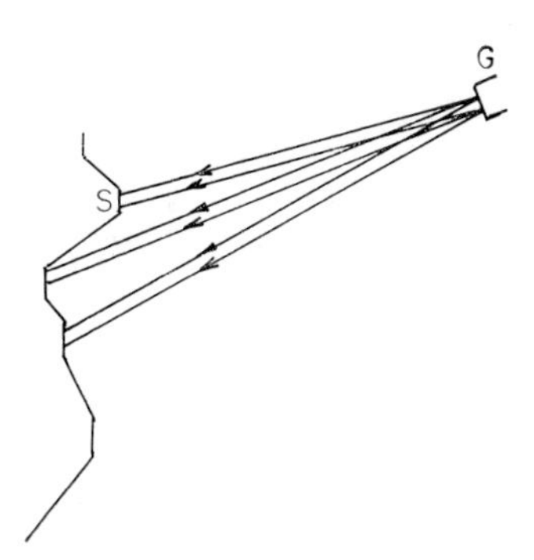

Fig. 5.13. Principle of the scanning microscope: if the incident beam of radiation from G, the gun, is near-parallel the scattering from the object is not affected by the variations in height of the specimen S.

about that one point. It does not matter whether the points scanned are all in one plane or not provided that the pencil of radiation is a parallel one. Fig. 5.13 is an attempt to illustrate this point. We shall consider some more of the practical consequences in the various sections of the final chapter.

5.7. *Conclusion*

In the five chapters so far, we have tried to lay the foundations for imaging theory. We have shown that in all imaging processes there are the two stages, the first being either scattering or radiation and the second recombination or focusing, and we have seen that this basic view of image formation applies to all kinds of radiation. We have seen that the scattering process itself can be studied in detail and sometimes indeed is the only aspect that can be studied. We saw for example that for X-ray studies of crystals it is impossible to complete the recombination process experimentally and we have to resort to alternative techniques. We have also seen that there are various different ways of performing the recombination stage, some of which, such as using a lens, enable us to do the whole thing simultaneously whereas others, such as scanning, involve point-by-point reconstruction. Finally we saw that all image-forming systems impose their own characteristics on the image that is produced and if we are to interpret our images properly we really should know as much as possible about our image-forming system. On the basis of all these ideas we shall now spend some time looking at a variety of practical image-forming systems to see how the principles already established work out in practice.

6. Applications and results

> " Merely corroborative detail, intended to give artistic verisimilitude to an otherwise bald and unconvincing narrative ".
>
> W. S. Gilbert
> *The Mikado*, Act II

6.1. *Introduction*

I have tried in the earlier chapters of this book to demonstrate that every system used to form an image conforms to the same series of basic physical principles. The illustrations I have used in the course of the book have been fairly general and there may have been numerous occasions when you wished for a little more detail or elaboration. In this last chapter, I propose to give practical details and descriptions for a considerable number of image-forming systems—and indeed in some cases systems that might not normally be considered as image-forming. I hope that they will not only prove interesting for their own sake but will also help you to see how the general principles developed in Chapters 1 to 5 work out in real life. It has always been one of my firm beliefs that physics should not be presented merely as a series of abstractions and generalizations. Though these clearly have a place, it is of great importance that they should be set in the context of real applications. This chapter therefore represents my attempt to be true to this belief. Inevitably it will be somewhat disjointed because it is really a collection of separate short descriptions. It also represents a personal selection which is far from being complete.

6.2. *The eye*

The eye is, of course, the most familiar of all image-forming systems and, indeed, we started the discussions in Chapter 1 with a consideration of the process of seeing. Fig. 6.1 is a standard diagram of a cross-section of a human eye. From our standpoint, the eye is a device which takes a sample of the pattern of light radiated or scattered by objects around us and, by means of a combination of phase adjusters, produces a representation on the retina which is then converted into the

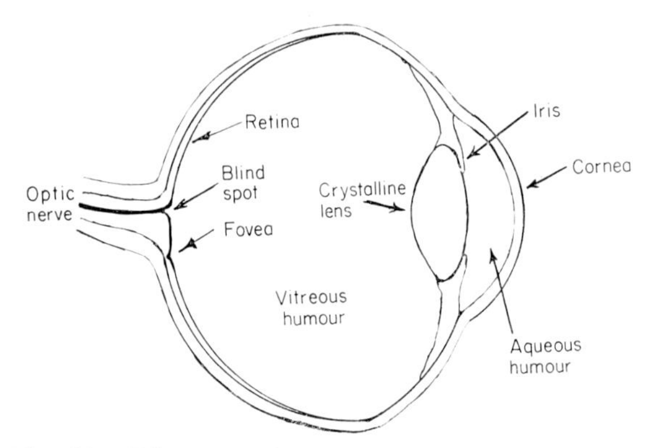

Fig. 6.1. Diagrammatic longitudinal section of an eye.

sensation of vision through a chain of electrical impulses to the brain. The phase adjusters are the combination of the curved surfaces of the cornea, the ‘ humours ’, or liquids which fill the eye, and the crystalline lens itself. The latter can be changed in shape by the ciliary muscle to vary the adjustment and hence to focus on either near or far objects. This process, known as accommodation, usually deteriorates with age.

An ordinary converging lens with spherical surfaces suffers, among other things, from the defect known as spherical aberration. The centre zone of the lens near to the axis has *less* phase-adjusting effect (By this I mean effect on the curvature of the incident wave, not the absolute phase shift which, of course, is greater at the centre where the lens is thicker) than the zones nearer the periphery, and the focal length for the outer parts is shorter than that for the inner. This can be demonstrated by forming an image of a distant lamp by a simple lens covered with a card with a small hole allowing light to pass only through the central zone of the lens. If the card is now replaced with one having an annular hole allowing light to pass only through the outer edges, refocusing is necessary and the outer part is easily shown to have a shorter focal length. The experiment works most effectively with a near-monochromatic source and a short focus lens.

The combined lenses of the eye also exhibit spherical aberration, but it is difficult to demonstrate this convincingly on a live eye because the effect is masked by other effects. In particular the increased depth of focus experienced when a small aperture is placed in front of the eye (discussed in the next section in connection with cameras) and the automatic focusing action of the eye lens create complications.

The eye also suffers from chromatic aberration and this can be demonstrated by observing a distant bright source of light through a piece of deep violet glass or gelatine which cuts out most of the centre region of the spectrum and passes only the red at one end and the violet at the other. Your eye will usually focus on the red image and this will be surrounded by a blurred violet image. Opticians sometimes use the effect as a ' fine adjustment ' in sight testing. A test chart is illuminated partly in red light, partly in green and partly in blue. When the patient sees the green part clearly and the red and blue parts are equally blurred, the correcting lenses are of the optimum focal length.

The final feature that is important from out point of view—though of course a whole book could be written about the eye and its quite remarkable properties and powers—is the beautiful economy of design of the eye.

If the eye is focused on a point source of light a long way away, we have exactly the conditions for producing Fraunhofer diffraction. The pupil of the eye will then give rise to an Airy disc around the image of the point source. The same effect will, of course, occur whatever the object being viewed; each part of the object will be reproduced on the retina as an Airy disc (compare fig. 5.3) and the image on the retina is a convolution of a true image and the Airy disc corresponding to the pupil.

Now the pupil varies in size from 1 to 6 mm according to the level of illumination. If we take 2 mm as the diameter in average daylight conditions, the angular diameter of the central disc on the retina will be just over 2 minutes of arc for a wavelength (measured inside the eye) of 5×10^{-7} m. The focal length of an average eye is about $1{\cdot}5 \times 10^{-2}$ m and so the kind of detail which is just on the limit of resolution is about 4×10^{-6} m. At the most sensitive area of the retina, the density of detecting cells is such that they are about 2×10^{-6} m apart. In other words the limitation on resolution provided by the structure of the retina is just a little better than the average limit set by diffraction at the pupil, a remarkable example of the economy arrived at during the process of evolution.

6.3. *The camera*

Many years ago when I was a young schoolboy I became interested in photography and a man I knew who was a keen amateur photographer offered to show me how to develop films and to make prints. He had a dark room fixed up in a cellar and I can still remember the astonishment I felt when he put the light out and there on the white-washed wall was an inverted image of a tree with its branches moving in the breeze—all in colour! My friend was annoyed and searched for the

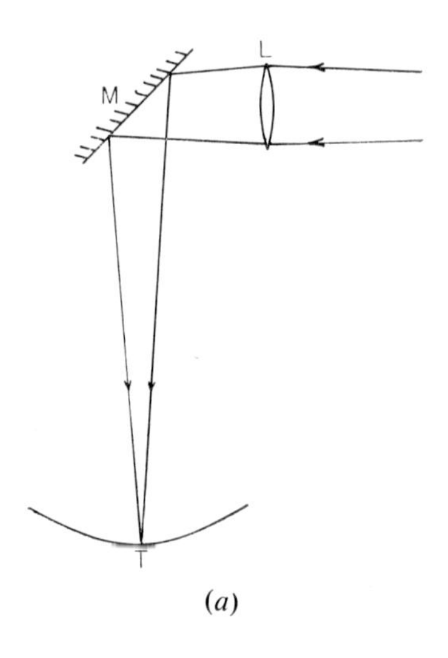

(*a*)

(*b*)

Fig. 6.2 (*cont. opposite*).

(*c*)

Fig. 6.2. (*a*) Principle of the Camera Obscura. (*b*) The Camera Obscura at Dumfries. (*c*) Inside the Camera Obscura at Dumfries. (Photo by David Hope, Dumfries.)

hole in the blind covering the window that was producing this unwanted image. It was of course a primitive pin-hole camera. I think the fascination that images have always held for me dates from that moment. Sometime later I visited a sea-side resort where a small hut on the cliffs advertised a ' Camera Obscura '. I duly paid my penny (old type) admission and was again enthralled. (Remember that this was before the days of television or colour films). There on the white table in the middle of the hut was displayed in full colour and, of course, complete with movement, a panorama of the beach and sea front. In this case a large lens in the roof was forming the image with a plane mirror to throw the image down on to the table. The system is just an enlarged version of the eye—though without the remarkable crystalline lens. Focusing was done simply by adjusting the position of the lens relative to the mirror and hence to the table (fig. 6.2 (*a*)).

During a recent holiday in Dumfries, I was delighted to find that the Camera Obscura at the Burgh Museum is still operating. It was set up in 1835 at a cost of £27.10*s*. It is believed to be the oldest Camera Obscura in the world that is still operational. The Old Mill which houses it can be seen in fig. 6.2 (*b*) and a group of visitors viewing the image on a concave table can be seen in fig. 6.2 (*c*).

The camera is a portable version of this device and it is important to remember that the camera focused on infinity is also a device for producing Fraunhofer diffraction patterns and that the image will be a convolution of the true image with the Airy disc of the lens. When focused on any other distance, the effect will be similar, as then the Fresnel pattern will resemble the Airy disc. A typical 35 mm camera (i.e. camera taking pictures on film 35 mm wide) might have a lens of focal length 5×10^{-2} m. The diameters of camera lenses are specified by so-called F numbers. This number is defined as the ratio of the focal length (f) to the effective diameter of the lens (d). Thus an F number of 8 means that the focal length is 8 times the effective diameter of the lens (I use the term ' effective ' because in a complex lens with many components it may not correspond to any *actual* component diameter but is the diameter of a thin single lens that would be equivalent). Now the magnification—if the lens is set to focus on a fairly distant object—is approximately the ratio of focal length to the distance away from the object (l) and so the *area* of the image of a particular object is proportional to f^2/l^2. The amount of light entering the lens is proportional to its area which in turn is proportional to d^2. Thus the image intensity (the amount of light per unit area) is proportional to $d^2 \div f^2/l^2$ which is simply l^2/F^2 and so for a given object distance the intensity of the image is proportional to $1/F^2$. Thus, if we choose an F number twice as large, the intensity goes down by a factor of 4. (Thus the exposure at $F16$ is 4 times that needed at $F8$.) Fig. 6.3 shows how lenses of different diameters and focal lengths can have the same F numbers.

The typical camera with a lens of 5×10^{-2} m focal length operating at $F2{\cdot}5$, would have a lens diameter at 2×10^{-2} m, and would produce

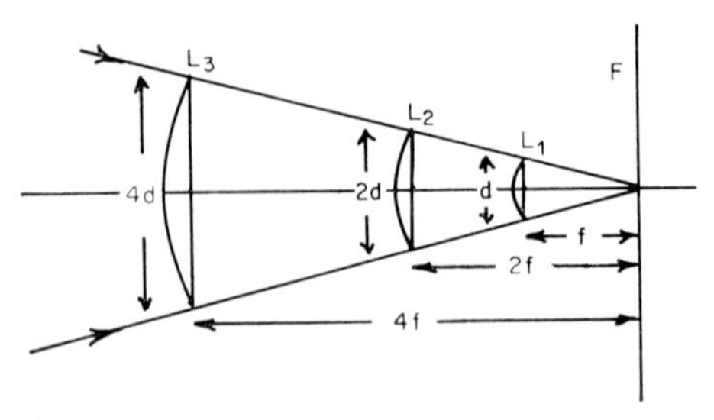

Fig. 6.3. Three lenses of different focal length but with the same F number.

a central disc with light of wavelength 5×10^{-7} m of angular diameter about 12 seconds of arc. Its diameter on the film would be about 3×10^{-6} m (=3 micrometres). Clearly this is not going to produce any serious problems unless the negative is to be enlarged to a very large size. The problem of the size of the individual grains of silver in the photographic image is likely to become significant before the lack of resolution of the lens unless very special fine-grain films are used. Fig. 6.4 shows examples of very highly enlarged negatives with both very fast (large grain) and very slow (fine grain) films.

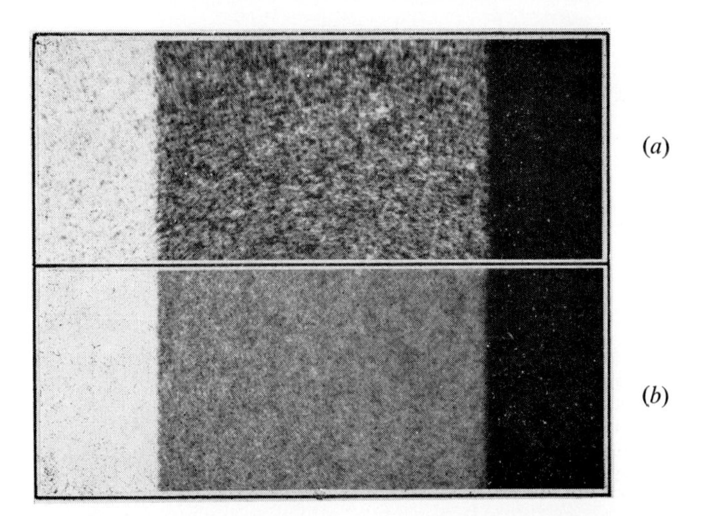

Fig. 6.4. (*a*) Micrograph $\times 200$ of white, grey and black lines photographed with the fastest film available for amateur photography. (*b*) Micrograph $\times 200$ of the same white, grey and black lines as for (*a*) but photographed on a very fine grain film for amateur use.

There are many fascinating problems that we could discuss in relation to cameras, but I shall pick out just two which link very closely with topics discussed in earlier chapters. The first concerns the depth-of-focus problem (see Section 5.6). The series of photographs of fig. 5.12 illustrates clearly the problem of reproducing a three-dimensional object on a two-dimensional film. In Chapter 5 we saw how scanning could overcome the difficulty and in section 6.9 we shall consider it in more detail.

There is, however, another important point to be made, and that is that the depth of focus is dependent on the lens aperture. This is most easily understood by a purely geometrical approach in the first instance. Fig. 6.5 (*a*) shows a convergent wave coming to a point

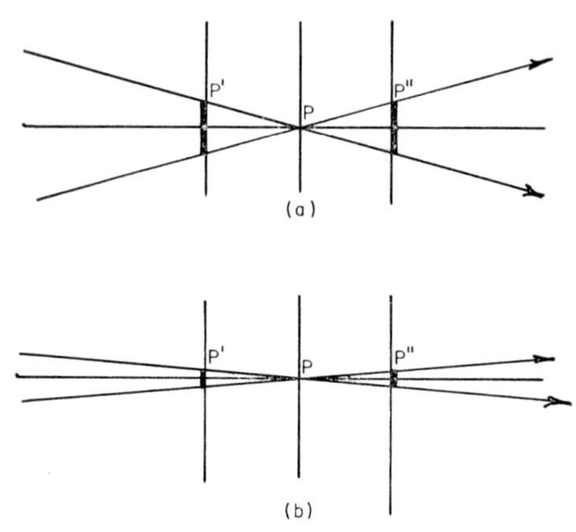

Fig. 6.5. Size of ‘ out-of-focus ’ patch (*a*) for lens with large *F* number. (*b*) for lens with much smaller *F* number.

focus at P and then diverging. At P′ and P″ the point would become a patch of a diameter corresponding to the diameter of the cone at these points. If we now consider fig. 6.5 (*b*) we have a converging wave of solid angle such as would be produced by a lens of lower *F* number. Now if we move to P′ or P″ the diameter of the patch is much less than in fig. 6.5 (*a*).

Suppose we are photographing a three-dimensional scene and the lens is adjusted so that a particular plane in the scene is focused sharply on the film. As in earlier examples we can imagine the scene to be made up of mathematical points and, when the image is sharply focused, each of these (if we ignore the Airy disc problem) is reproduced as a mathematical point. If we now transfer our attention to a plane at some other distance from the lens which focuses at a point in front of or behind the film, then each point on the object will be reproduced as a circle as at P″ in fig. 6.5 (*a*) or (*b*). The image will now be a convolution of the true image with the circular patch.

It should now be clear from figs. 6.5 (*a*) and (*b*) that for a given plane on the object the patch size will be greater, and the blurring of the image consequently also greater, the larger the aperture of the lens; or putting it the other way round, if the eye can tolerate blurring up to a certain patch diameter, planes in the object of greater ‘ front-to-back ’ depth will appear ‘ in focus ’ if the aperture is smaller than if it is larger. Fig. 6.6 shows a series of photographs of the same object from the same position and with the same lens but with varying apertures. (The exposures were adjusted to compensate for the varying amounts of light admitted.)

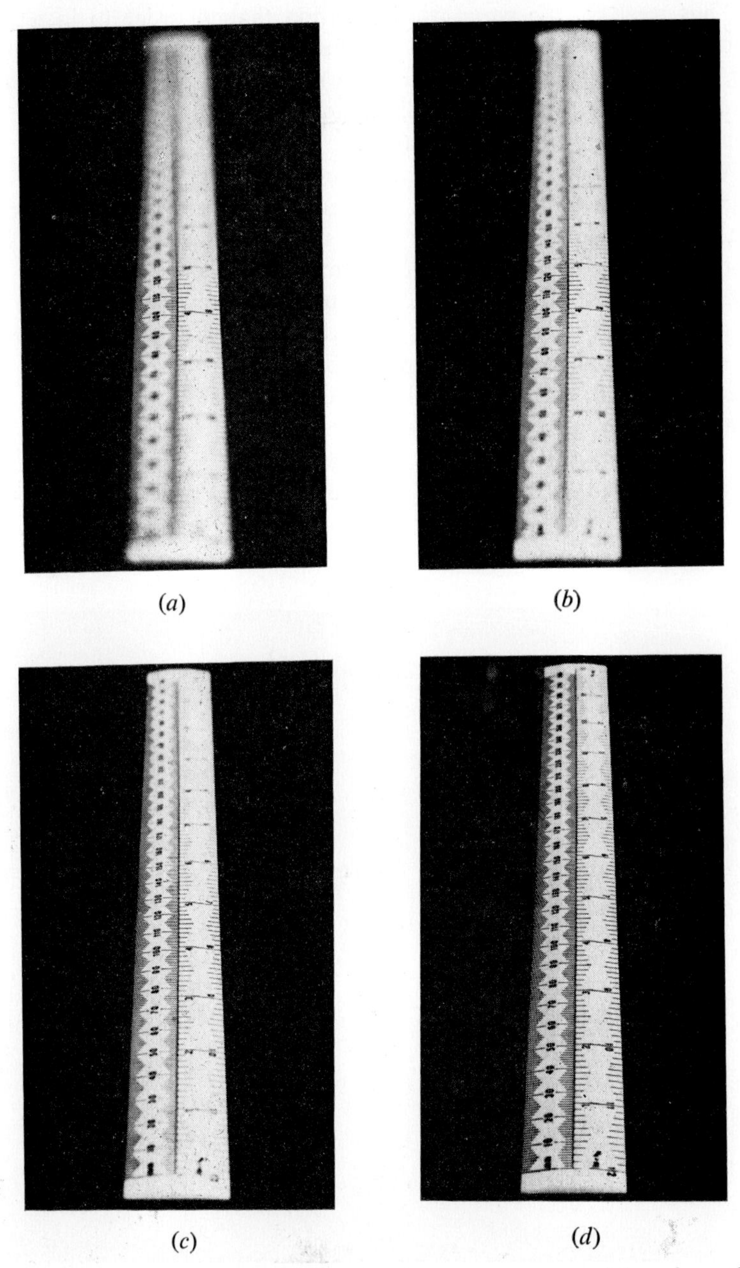

(a) (b) (c) (d)

Fig. 6.6. Photographs of a ruler with the same camera and lens from the same position and in each case focused on the 100 mm mark. The exposures were adjusted to give similar contrast. (*a*) *F* 1·8; (*b*) *F* 4; (*c*) *F* 8; (*d*) *F* 16.

Thus we have the paradoxical situation that in order to increase depth of field we need to use a *small* aperture but in order to avoid diffraction problems we need a *large* aperture. In practice, as has already been mentioned, the diffraction problem is rarely of great consequence in normal photography.

The final point I want to mention here, because it links well with our other discussions, concerns trick effects that are sometimes used with cameras—particularly cine cameras. We have already used the notion that a camera focused on a point source is a suitable device for obtaining Fraunhofer diffraction patterns, and indeed we saw in section 5.1 that a point source would produce the Airy disc of a lens aperture. If some kind of mask made up of opaque and transparent areas is placed over the lens, then the lens aperture is being multiplied by a function represented by the pattern of the mask. The image must then consist of the *convolution* of the true image with the diffraction pattern of the patterned mask. Fig. 6.7 (*a*) shows what happens when a scene containing several bright lights is photographed with fine silk gauze over the lens. Since the centre spot of the diffraction pattern of the

(*a*)
Fig. 6.7 (*cont. opposite*).

(*b*)

(*c*)

Fig. 6.7. (*a*) Scene photographed with gauze over lens of camera. (*b*) As for (*a*) but with gauze rotated. (*c*) Close up view of gauze.

gauze is so much brighter than everything else, the convolution with the background scene reproduces the scene fairly well but where there is a bright light the full diffraction pattern stands out. Rotating the gauze in its own plane leads to a rotation of *each* of the diffraction patterns about its own centre, as shown in fig. 6.7 (*b*). This is a very good additional illustration of the convolution–multiplication property.

6.4. *Optical telescopes*

Telescopes are often presented as devices for producing enlarged images of distant objects but, as will have been realized from earlier chapters, the enlargement is only part of the story; the increased light-gathering power above that of the pupil of the eye can be every bit as important. The usual diagrams of the two simplest telescope systems with lenses emphasize the light-gathering feature more than the magnification. Fig. 6.8 (*a*) shows the Galilean telescope (sometimes called a terrestrial telescope as it does not invert the image) and 6.8 (*b*) shows the Keplerian telescope (sometimes called the astronomical telescope because it inverts the image but this is of no consequence in viewing stars). In each a parallel beam of considerable cross-section enters the objective and emerges still parallel but very considerably reduced in cross-sectional area; the intensity of the image is thus considerably increased. Simple telescopes of this type do not give images that are free from aberrations, and many variations using compound lenses for the objective to correct for both chromatic and spherical aberrations and various other errors are known; a great variety of different eye-pieces, most of which correct at least to some extent various aberrations, may also be incorporated.

Because of the difficulty of casting very large lens blanks, the problem of supporting them without interfering with the optical properties and the complexity of correcting the various aberrations when the aperture

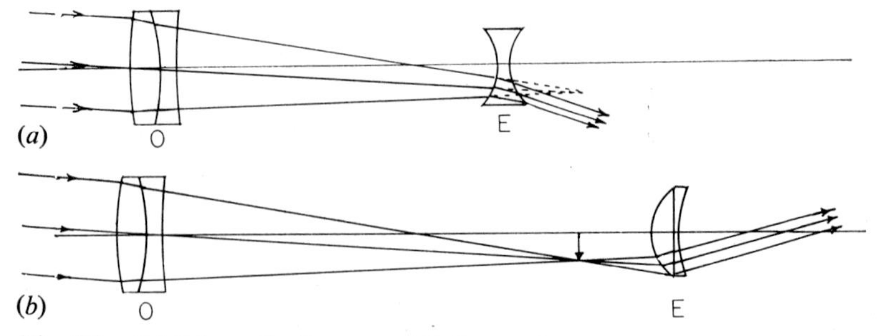

Fig 6.8. (*a*) Schematic diagram of Galilean telescope in normal adjustment. (*b*) Schematic diagram of Keplerian telescope in normal adjustment (often simply described as ' the astronomical telescope ').

is large, most really large telescopes are made using a mirror as the objective. A spherical mirror reflects all wavelengths in the same way and hence is inherently free from chromatic aberration. If the mirror is made paraboloidal instead of spherical, the spherical aberration problem disappears and a parallel beam of light covering the whole aperture of a large mirror can be brought to a point focus.

Fig. 6.9 shows the Newtonian telescope arrangement. The paraboloid collects the light and the image is reflected through the side wall by a small mirror, known as the ' diagonal ', to an eyepiece.

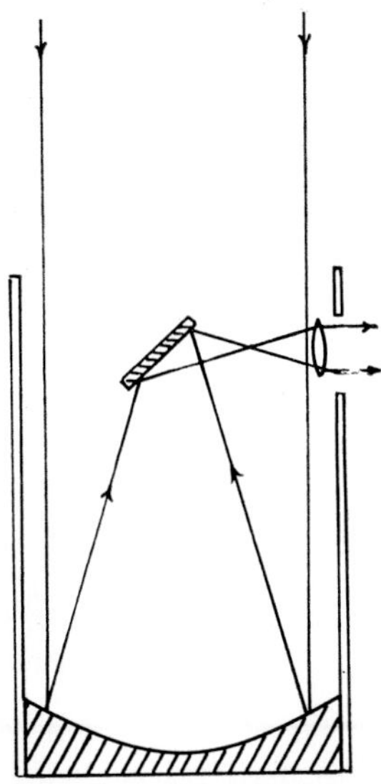

Fig. 6.9. Schematic diagram of Newtonian reflecting telescope in normal adjustment.

What is the effect on the image of interposing this mirror and its support? We can approach this problem using our knowledge of diffraction and convolution. The system gives, for each object point, a figure which is the diffraction pattern of the aperture; the aperture, in this case, is a large circular hole with a relatively small opaque spot at the centre. Optically, the combination may be thought of as a small circular hole subtracted from a larger one! Each hole has as its diffraction pattern an Airy disc and we need to subtract the one from the other. Fig. 6.10 (*a*) shows the distribution in amplitude of the Airy disc from the large aperture and (*b*) shows that from the small aperture assumed to be one tenth the diameter of the large one; (*c*) shows their difference. It is clear that, although the fringe pattern is modified slightly, the predominant effect—the central peak—is very little

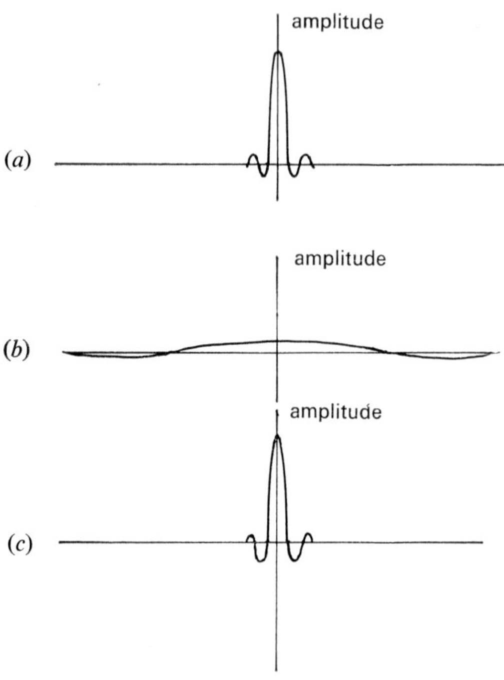

Fig. 6.10. (*a*) Amplitude distribution in Airy disc pattern of full aperture. (*b*) Amplitude distribution in Airy disc pattern of an aperture of the same size and shape as the diagonal mirror. (*c*) (*a*) minus (*b*), the amplitude distribution in the diffraction pattern of the annulus which is the effective aperture of the Newtonian telescope.

different from that due to the large aperture on its own. And this really answers the question: the convolution of any image with the function of 6.10 (*a*) will not differ very much from its convolution with 6.10 (*c*) and hence the diagonal does not upset the image.

Again there are many variants on this system, including the Hale type of telescope in which the observer himself is seated inside the telescope near the focal point—though clearly this can only apply to pretty large instruments such as the 200 inch Mount Palomar telescope. Whatever kind of telescope is being used, however, the basic function is to pick up as big a fraction as possible of the scattering pattern produced by the object under examination and to recombine it into an image; a major part of the problem of telescope design is to ensure that this recombination is done with as little error as possible. One of the principle points of interest, therefore, is the method of testing the spherical or paraboloidal mirrors developed for telescopes, partly because the successful operation of telescopes depends critically on the success of this testing, and partly because the test itself relates very directly to our discussions on image formation in the earlier chapters.

Let us first consider a method of testing a spherical mirror and let us further assume to begin with that our mirror is perfect. We set up a tiny point source of light and, placing it next to one eye, move both the eye and the point source around until we have the source and eye very close to the centre of curvature of the mirror (fig. 6.11 (*a*)). In terms of image formation the mirror is being illuminated and its scattering pattern is the returned image of the point source—the Airy disc of the aperture. The eye recombines this to give an image which shows completely uniform brightness over the surface of the mirror; it is this uniform brightness that signifies to the observer that his eye and the source are close to the centre of curvature.

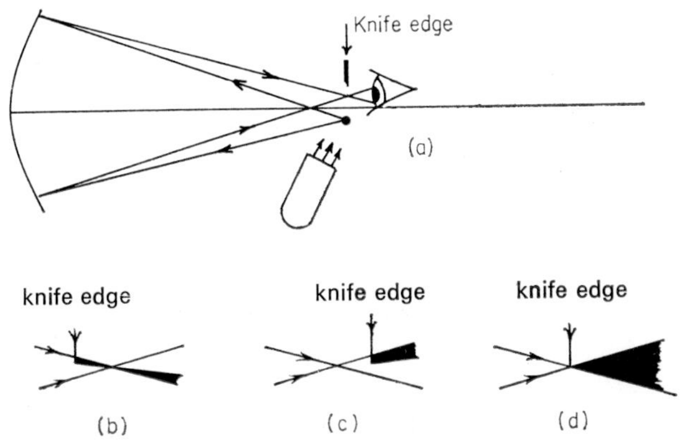

Fig. 6.11. The Foucault 'knife-edge' test for spherical mirror. (*a*) The experimental arrangement. (*b*) Knife edge nearer to mirror. (*c*) Knife edge nearer to eye. (*d*) Knife edge at centre of curvature.

Now a knife edge, often a razor blade, is introduced in front of the eye and is passed (say) from left to right across the field of view. Simple geometry (fig. 6.11) shows that, if in fact the blade is nearer to the mirror than its centre of curvature, the eye will be conscious of a shadow moving from right to left (*b*). If the blade is nearer to the eye than the centre of curvature of the mirror, the shadow will move with the blade from left to right (*c*). If the blade is *exactly* at the centre of curvature the whole aperture of the mirror will appear to go dark instantly (*d*).

Now let us suppose that there are some small errors in the mirror—perhaps odd scratches, patches of dirt or departures from a perfectly spherical shape. Without the blade in position, unless the effects are really gross, they will not be seen by the eye. However, if the blade is brought across *at* the centre of curvature until the illumination of the whole mirror is *just* extinguished, all the defects will suddenly appear quite clearly. Why is this?

The defects can be thought of as objects across the aperture of a system producing an approximation to the Fraunhofer diffraction pattern and so the image formed near the eye will no longer be a simple Airy disc but may be quite a complex pattern. If the defects are smaller, as is inevitable, than the aperture of the mirror, then, because of the now familiar reciprocal relationship between objects and their diffraction patterns, the diffraction pattern will be *larger* than the Airy disc. Consequently when the blade has advanced to cut off the Airy disc and hence the main illumination of the mirror, the larger parts of the diffraction pattern will still be available to the eye for recombination. Following Abbe's principle, the eye will recombine the pattern to form an image of something that would give just *that* as its diffraction pattern without the Airy disc; in other words it produces an image of the errors without the overpowering overall illumination of the whole surface. This test, known as the Foucault knife-edge test, forms the basis (though a great many complicated variations have been developed) of most methods of examining the perfection of mirrors. It is also very closely related to the Schlieren technique to be described in Section 6.6.

A paraboloidal mirror has to be tested in a slightly different way, with an auxiliary optically flat mirror. Fig. 6.12 shows the arrangement; the eye and source are now at the focus of the paraboloid, and hence a parallel beam is produced which is then returned by the flat and hence to the focal point. The effect of errors or imperfections can be understood in exactly the same way as for the spherical mirror. The provision of a suitable point source sometimes presents difficulty and an excellent solution is to use a highly polished steel ball to create a

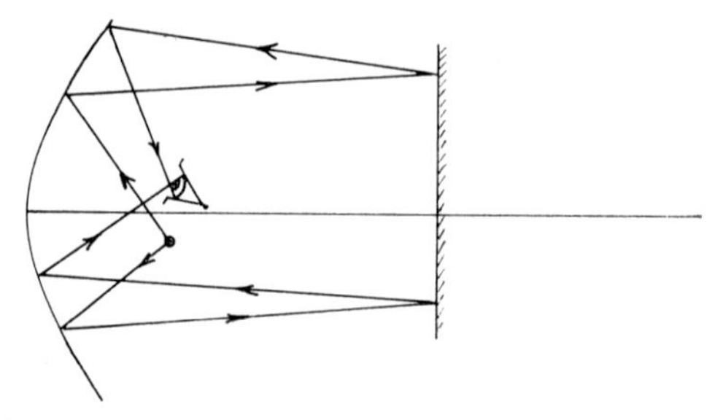

Fig. 6.12. Adaptation of knife-edge test for paraboloidal mirror.

small image of a bright lamp placed some distance away to one side. Such a source—i.e. the steel ball—can be placed very closely to the eye without discomfort.

The process of shaping or testing a paraboloidal mirror is a fascinating one but is also quite difficult to achieve with high precision, and consequently a number of alternatives have been developed. Perhaps the most famous of these is the Schmidt system, which is widely used for medium size astronomical telescopes for photographic purposes (in the range up to 2 m diameter). Again it is picked out here because it provides a beautiful illustration of some of the principles developed in earlier chapters.

A spherical surface is relatively easy to produce. Experience shows that if two pieces of glass are rubbed together with grinding powder between and are frequently rotated relative to each other, one becomes concave and the other convex and both acquire spherical surfaces to quite a high degree of precision. Indeed spherical surfaces are the only possible ones (other than absolutely flat surfaces) which can slide over each other in such a way that all points remain in contact. If the surface were *other* than spherical the high points would be subject to grinding and the low points protected and the surfaces would gradually, and more or less automatically, become spherical.

The Schmidt system starts with a spherical mirror and then attempts to correct for the difference between the required paraboloidal surface and the given sphere. Figs. 6.13 (*a*) and (*b*) show the difference between the behaviour of a paraboloid and a sphere. The waves incident near the centre of the sphere come to a focus too far away from the mirror and those near the periphery come to a focus too near to the mirror. The Schmidt system involves adding a so-called corrector plate which is a sheet of transparent material of thickness which is different at different distances from the centre. The introduction of this plate effectively multiplies the amplitude distribution of the wave front by the transmission function of the plate. But since it is almost completely transparent the magnitude is unaffected and it is only the relative phases of the waves that are affected. The device can be thought of as introducing a ' formant ' to the system (see Section 5.4). A cross-section of a corrector plate is shown in fig. 6.13 (*c*). It can be seen that near the centre the added phase difference diminishes as we move out from the centre; the plate therefore acts like a converging lens and this in turn moves the focal point of these waves nearer to the mirror. Towards the periphery, the plate thickens again and so in the outer region it acts like an annular *diverging* lens and so moves the focal point of the peripheral waves away from the mirror. If the dimensions of the plate are carefully chosen, all the waves arrive at the same focal point and the combined plate and spherical mirror behave like a paraboloidal mirror. A fuller mathematical investigation shows that if the plate is placed in the plane of the centre of curvature of the mirror other aberrations are diminished as well and the system behaves better even than a paraboloid. Fig. 6.13 (*d*) is a view of a building with an ordinary camera; fig. 6.13 (*e*) is a photograph of a detail of the building from exactly the same position (90 m from the building) using a small (0·12 m) Schmidt system. The remarkable reproduction of detail testifies to the quality of the optical system.

Two final points in our consideration of optical telescopes will now be mentioned. Both involve restricting the aperture in certain ways in order to perform measurements. The first permits the measurement of the diameter of a star and is the so-called Michelson stellar interferometer. In its simplest form it consists of two slits, whose separation can be varied, placed over the aperture of the telescope. Following our usual practice we can think of this double slit as a function multiplying the aperture and hence instead of the usual Airy disc pattern, each star image will be convoluted with the diffraction pattern of the double slit. The diffraction pattern of the double slit is of course a normal Young's fringe pattern and so each star image now has fringes across it as well. If the spacing of the slits is increased the spacing of the fringes is *decreased* but the fringes will remain visible only if the illumination of the slits is *coherent*. Now the degree of coherence of the illumination of the telescope aperture depends on the size of the star source. The bigger the source the lower the coherence and so the more quickly will the fringes disappear as the slit spacing increases. A fairly simple mathematical relationship permits the separation of slits for which the fringes disappear to be correlated to the star diameter. The experiment is precisely the same in principle as that described in Section 2.8 for the measurement of spatial coherence.

For small stars, the aperture of the telescope may not be big enough to allow the slits to be moved sufficiently far apart. A framework carrying moving mirrors (fig. 6.14) can be added to the telescope to increase the effective aperture for the purposes of the measurement.

The other modification of the aperture that is of interest is the technique known as apodization. A telescope of given diameter cannot

Opposite:

Fig. 6.13. (*a*) Parallel beam incident on paraboloidal mirror. (*b*) Parallel beam incident on spherical mirror. (*c*) Schmidt plate which will make (*b*) behave as (*a*). (*d*) View of building with ordinary camera. (*e*) View from same position with 0·12 m Schmidt camera.

(a)
(b)
(c)
(d)
(e)

resolve two stars closer together than a certain angular separation and it would seem that the only way to improve the resolution would be to *increase* the aperture—usually an impractical solution. Paradoxically it can be shown that by blocking up part of the aperture some improvement can result! A circular opaque disc is added to leave an annular ring of the lens or mirror as the only part contributing to the image.

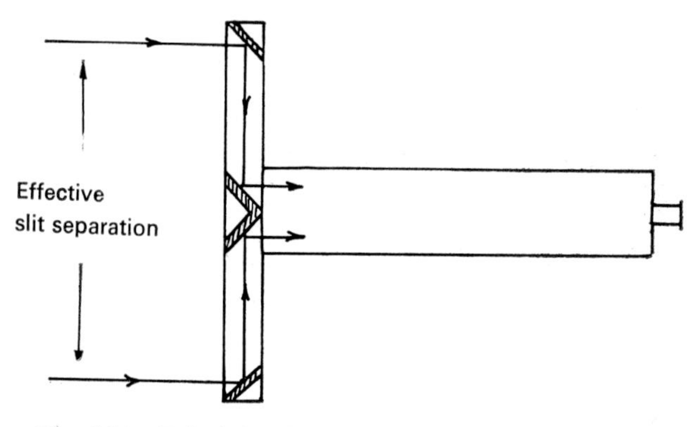

Fig. 6.14. Principle of the Michelson stellar interferometer.

We can again understand how it works if we remember that the diffraction pattern of this annular ring will be reproduced by convolution at each star image. Fig. 6.15 is a plot of the Airy discs of the whole aperture of the telescope and of an aperture the size of the opaque disc. The diffraction pattern of the annulus is the difference of the two, which is shown on the same figure as the thick line. The important point to notice is that the central peak of the diffraction pattern of the annulus has a *smaller* diameter than that of the full aperture. Hence the resolution will be increased.

But this seems like something for nothing! We seem to be gaining information by throwing information away. The resolution of the paradox is also contained in fig. 6.15 (and in fig. 2.12 (f) which shows the actual diffraction pattern of an annulus). You will notice that the rings surrounding the central peak for the pattern of the annulus are much more intense than those for the plain aperture. The whole image is therefore very messy and much more muddled with subsidiary rings and detail. Nevertheless if we take a photograph with the full aperture first and know where a particular star suspected of being a doublet is located, a subsequent photograph with the apodizing disc in place may settle the question of whether or not it is a doublet while

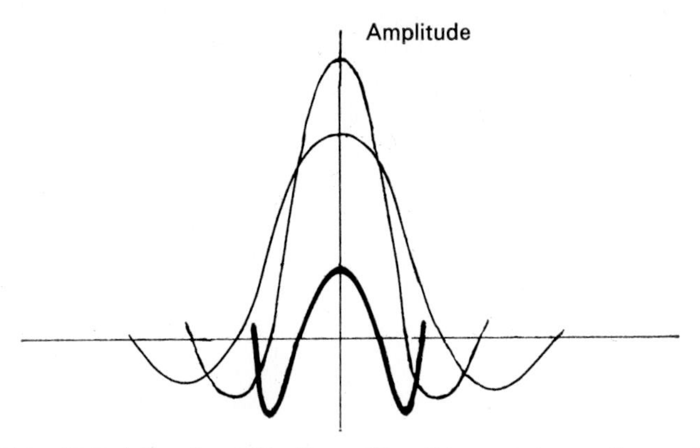

Fig. 6.15. Principle of apodization. The thin curves are the amplitude distributions in the Airy disc patterns of the whole aperture and of an aperture of the size of the opaque disc. The thick line is their difference and shows that the central disc is much smaller than that for the whole aperture—but the first ring is greater in amplitude.

giving many other spurious effects which we know from the first photograph may be ignored.

6.5. *Radio telescopes*

In the early 1930s the attention of radio engineers became fixed on the origins of some of the curious background noises that were picked up on radio receivers and which did not seem to have an origin in known terrestrial transmitters. Some of these noises were quickly traced to the electrical discharges in thunder storms and the first-ever radio telescope was probably a rotatable aerial designed to locate the storm centre giving rise to ' atmospherics ' as the cracklings were called. In the course of this work it was found that there were variations in background noise which had a regular period of about a day. Careful studies then showed that the period of variation was about 4 minutes *less* than a solar day of 24 hours—it was in fact tied to the *sidereal* day, that is the day measured in terms of the apparent rotation of a particular *star* through a complete cycle round the Earth rather than of the Sun. This careful observation showed that the source of the background noise in radio transmissions must be associated with a particular region of space.

Since that time an enormous number of developments have occurred and many famous radio telescopes—such as the 250 ft steerable telescope at Jodrell Bank—have been built. Fig. 6.16 shows this impressive structure which makes a striking object on the sky line as one drives north through Cheshire on the M6 motorway.

The physical principle of the radio telescope is exactly the same as that of the optical one. The common form is a large reflecting dish which behaves in exactly the same way as the paraboloidal mirror in a reflecting telescope. This dish samples the scattering pattern, which is usually radiated by sources rather than being scattered by them, and the resolving power is calculated in terms of the size of the Airy disc of the aperture.

Fig. 6.16. General view of 250 ft steerable radio telescope at the Nuffield Radio-Astronomy Laboratory, Jodrell Bank, Cheshire. By permission of the University of Manchester Nuffield Radio Astronomy Laboratories.

One important wavelength of radiation coming from interstellar matter is at 0·21 m. This is about 3×10^5 larger than wavelengths in the visible region, and so to match the resolving power of a small amateur reflecting telescope for visual astronomy of say 0·15 m diameter we should need a *radio* telescope dish 45 km in diameter!

This is not in fact *quite* as ridiculous as it sounds for the following reason. A radio telescope does not produce a picture of a radio source directly as does an optical telescope. With the dish fixed at one position and pointing in one direction the antenna placed at the focal point of the dish records a particular level of signal. Now of course the interest may be in the *time* variations of this signal—as was the case of the early observations mentioned above. But if the spatial distribution is of interest the telescope must be made to change its direction or to move its position on some kind of railway track. What is then explored is the *radiation* pattern of the radio source at the Earth's surface and from the appropriate measurements the characteristics and shape of the source can be deduced. In particular if two small telescopes are set up, one fixed and the other moving along a straight railway track, it is not only possible to compare the intensity at different points in the diffraction pattern but also comparisons in the relative phase can be made. This is the advantage of working with frequencies of only about 10^9 Hz involving time measurements in nanoseconds which are perfectly possible. (Visible light, you will remember, has frequencies in the 10^{14} Hz region.) Thus a fairly accurate record of the diffraction pattern along this straight line can be built up and from this a precise picture of the radio source can be produced. There is no reason why such a track should not be 45 km long and this relatively simple system would give a resolution along this particular line comparable with that of 45 km diameter telescope!

There are many variations on this theme. Some telescopes are kept fixed to the Earth's surface and merely make use of the Earth's rotation in space to trace out the radiation patterns. Others use a large number of small dishes dotted about over a large area to sample the radiation pattern and hence to build up the required pictures.

6.6. *Schlieren techniques*

In our discussion of the Foucault knife-edge method of testing telescope mirrors, the point was made that various kinds of deviation from a perfect spherical mirror such as dust, scratches, grease, etc., as well as actual errors in the shape are revealed. It is not difficult to see that, in a similar way, any departures from uniformity in the optical behaviour of the medium in between the mirror and the focal point would have similar effects. Thus if we set up the system for the knife-edge test with an absolutely perfect and clean spherical mirror and adjust the knife edge so that the field *just* goes dark we then have a sensitive system for detecting variations in the refractive index of the air between the mirrors and the focal point. Fig. 6.17 shows the use of the system to study the initiation of an edge-tone (such as occurs in an organ pipe) using a rising column of hot air.

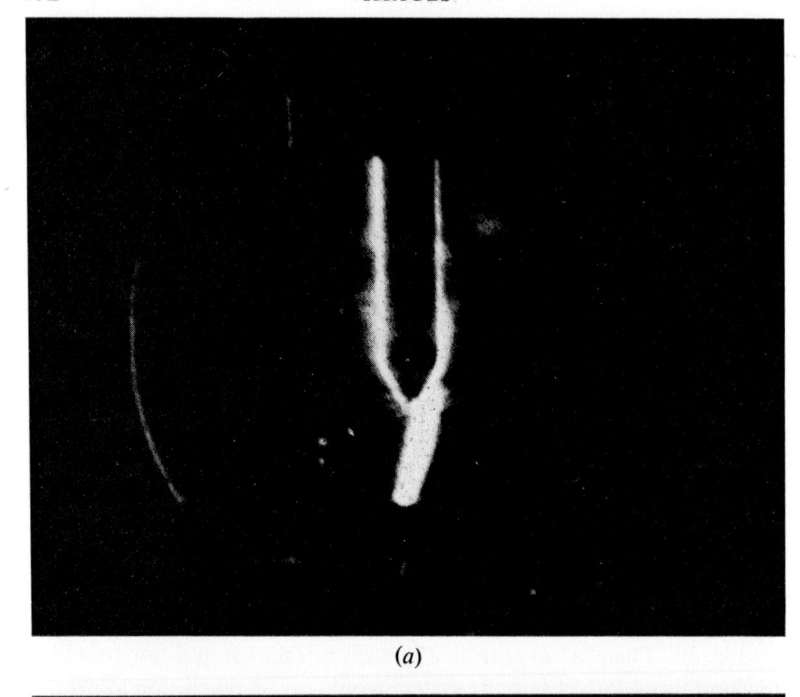

(*a*)

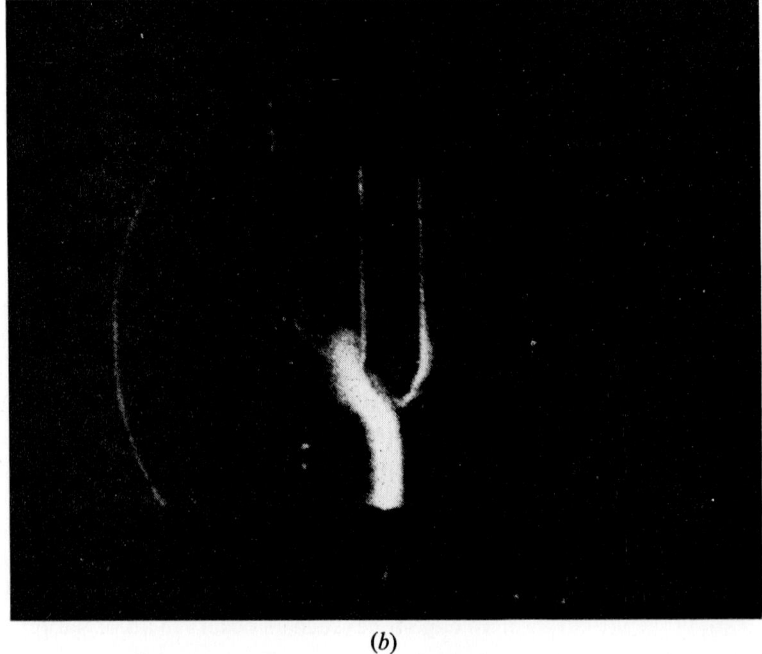

(*b*)

Fig. 6.17. (*a*) and (*b*) Successive stages in the formation of an ' edge ' tone by a column of air rising and incident on an edge. The changes in density of the air are revealed by the Schlieren technique.

This kind of system was first introduced for examining the perfection of slabs of optical glass. The problem in producing optical glass is that the melting point is close to the point at which the material of the lining of the ladle dissolves in the glass: if the temperature is just too low, the glass is very viscous and it is difficult to remove tiny bubbles; if the temperature is just too high, constituents of the ladle lining may go into solution and produce regions of different refractive index. Since the glass is usually stirred before pouring these regions become streaks in the glass and the word *Schliere* is simply the German for ' streak '.

Perhaps the most spectacular application of the technique has been in studying the behaviour of model aircraft in wind tunnels where the Schlieren technique can be used to reveal the variations in refractive index of the air flowing over the aerofoil surface, which of course are related to the velocities and densities of the air. Special systems are needed; a common modification of the basic Foucault system is shown diagramatically in fig. 6.18. Two mirrors are used and again it can be seen that the system satisfies the conditions for producing an intensity distribution corresponding to the Fraunhofer diffraction pattern of whatever is put in the parallel beam. The knife-edge eliminates the bright Airy disc which would transform back to the uniformly bright field and the diffraction pattern of any fine details gets past the edge of the blade and recombines to form images of the disturbances in which we are interested. The question is sometimes raised that the knife-edge cuts out half the diffraction pattern of even the parts of the image in which we are interested. This is true and indeed the resultant appearance of the field is different from that which would result if we replace the knife-edge by an opaque dot exactly the

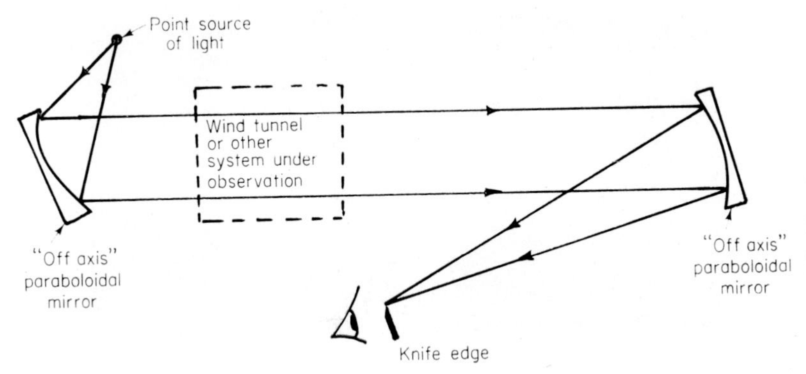

Fig 6.18. Diagram of a typical Schlieren system using two large mirrors.

size of the Airy disc. But the detail is clearly visible and the defects in the image are offset by the enormous increase in difficulty of locating the small opaque dot!

In a very beautiful and useful modification of the Schlieren technique, the knife-edge, which can be thought of as a strip of opaque material and a strip of transparent material with a straight line border, is replaced by a strip of red colour filter and a strip of green colour filter with a straight line border. The central Airy disc passes half through one and half through the other and hence when the recombination takes place the image of the aperture is a mixture of red and green light, that is yellow. But any detail in the field which produces diffraction effects in the green filter area will be reproduced in green and those giving diffraction detail in the red area will be reproduced in red. With care it is possible to relate the precise colour to the variation in refractive index of the medium and so the picture can be interpreted quantitatively.

6.7. *Image-processing*

You will remember that in Section 5.3, in which we discussed Abbe's theory of microscopic vision, it became clear that the final image (and the principle applies to any image-forming system, not just to microscopes) is not necessarily a reproduction of the object: rather it is a reproduction of a hypothetical object which would give a scattering pattern identical with the modified pattern which is actually allowed to proceed to the recombination stage. It thus becomes obvious that seeing is *not* believing, as by modifying the scattering pattern in different ways we can make the image almost anything we want.

Fig. 6.19 shows an example which was prepared using the laser diffractometer described in Section 2.6, with the adaptation to permit

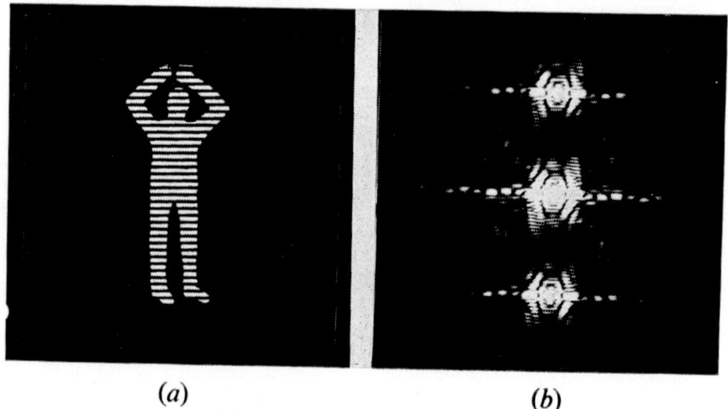

(*a*) (*b*)

Fig. 6.19 (*cont. opposite*).

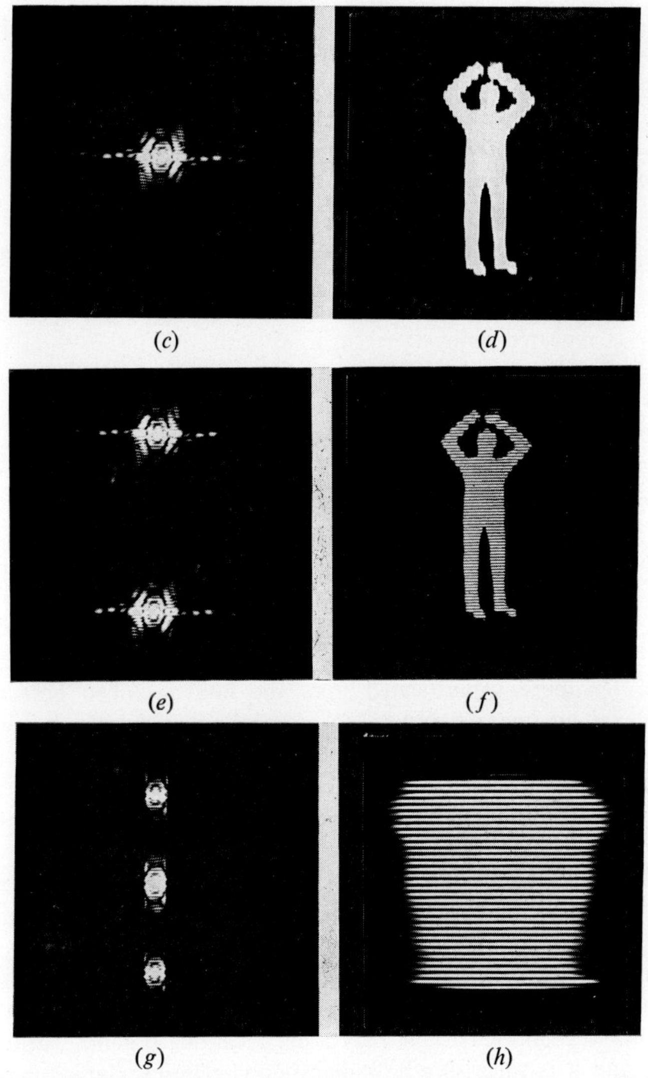

Fig. 6.19. (*a*) Object made up of transparent and opaque regions. (*b*) Centre region of diffraction pattern of (*a*) showing only 3 repeats in the vertical direction: in the full pattern there are several more repeats above and below those shown. (*c*) Central unit only of (*b*). (*d*) Recombination of (*c*). (*e*) Central unit of (*b*) and all others except the adjacent one above and below the centre eliminated. (*f*) Diffraction pattern of (*e*): note the spacing of the stripes relative to those in (*a*). (*g*) All units left in vertical direction but laterally the pattern is restricted to a very narrow vertical strip. (*h*) Recombination of (*g*). From *An Atlas of Optical Transforms*, by G. Harburn, C. A. Taylor and T. R. Welberry, by permission of G. Bell & Sons Ltd.

the diffraction pattern to be modified and then recombined; the schematic diagram of fig. 2.20 (*a*) showed the system. The object 6.19 (*a*) is an outline of a man crossed by stripes. This particular object is chosen because striped pictures commonly arise as a result of television transmissions, for example, from space-craft. The object can be thought of as the product of the man's outline and the set of stripes. Thus its diffraction pattern (6.19 (*b*)) will be the convolution of the diffraction pattern of the outline with that of the stripes; the latter is a set of points (the so-called ' orders of diffraction ') arranged in a vertical row and so the odd-shaped diffraction pattern of the man's outline is repeated many times in a vertical row (fig. 6.19 (*b*) shows only the 3 centre repeats). Fig. 6.19 (*c*) shows the central version on its own with the rest masked off; when recombined therefore there is no information about the stripes and they disappear (fig. 6.19 (*d*).) If the central pattern is eliminated and just the one on either side allowed to contribute (fig. 6.19 (*e*)), then since these two are twice as far apart, the stripes appear to be of half the width! (fig. 6.19 (*f*)). Finally, if all the orders are left in but the width of each is severely restricted (fig. 6.19 (*g*)) then the stripes remain but the outline is largely lost (fig. 6.19 (*h*)).

There are all kinds of practical applications of this technique which can be far more sophisticated than the one described here. For example, if an imaging system has some known defects, it may be possible to modify the scattering pattern in such a way as to compensate for these defects and so produce a better image.

A word of caution is necessary though. Image-processing techniques may make an image far more acceptable to the eye but they cannot actually increase the amount of information present. A crude illustration of this might be that if a television picture is produced of a man with a pimple on his chin and it so happens that the pimple is completely covered by a *dark* stripe and no trace of it is visible in the light stripe on either side, then no amount of image filtering will ever reproduce the pimple! This may seem facetious but the warning is serious and it is surprising how many good physicists are sometimes misled into believing that they can extract more information from processed pictures than is there before processing.

There are, of course, some systems in which processing can actually lead to the extraction of more information—but it is very important to think clearly about what is actually happening before reaching conclusions. For example, a television picture from a space vehicle may be badly affected by noise producing a strong ' snow-storm ' effect over the picture. If the picture being transmitted is a still one so that there is no change in the object over many frames it is quite possible to introduce electronic filtering based on comparisons between successive frames. Data which appear in two successive frames are preserved and any which appear in only one frame are eliminated. If this process is repeated over a period very great enhancement of picture quality is possible. We are in fact using one of the principles of focusing that was elucidated in the first chapter. We *know* that the picture we want is unchanging; we *know* that the noise that we do not want is changing in a random way. This knowledge enables us to separate the two.

6.8. *Optical microscopy*

I have used optical microscopy to illustrate numerous aspects of imaging theory already in the course of the book and, of course, a special section was devoted to the Abbe theory of microscopic vision (Section 5.3). There are, however, two special modifications of microscopes that have not so far been mentioned and which are splendid applications of imaging theory that I think merit inclusion. I have also included—largely because I am a great admirer of the quite remarkable work done by microscopists in the late ninteenth century—a historical reference that provides an interesting example of indirect recombination used in optical microscopy.

The first modification involves the technique known as dark field illumination. It is, in fact, very closely related to the Foucault knife-edge test and to the Schlieren techniques already described. In a normal microscope the object is illuminated by a converging cone of light from the illuminating system below the microscope stage. In the dark field system, an opaque disc of suitable size is placed in the centre of the illuminating system so that the cone of illumination becomes hollow. If the stop is of the right size, the field of view on looking down the microscope is entirely dark when no object is in place. When an object is introduced it can be seen outlined in light against the dark background. The system can best be understood by thinking of Abbe theory. In effect the stop has removed the central order but since, if there is no object, the central order is the only one present in the diffraction pattern, the recombination results in a blank field (fig. 6.20). Any object in the field however will create a diffraction pattern which will get past the opaque disc and recombine to give a picture of detail resembling a Schlieren pattern. The technique is particularly useful when highly transparent objects are being viewed. Differences in thickness and refractive index may be so small that in

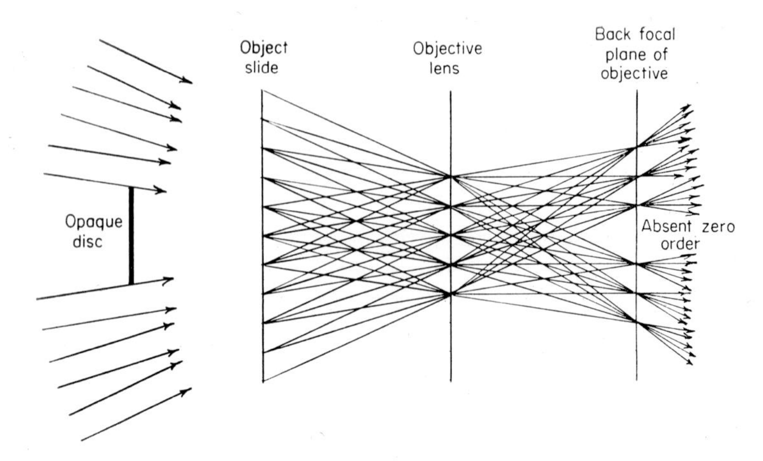

Fig. 6.20. Illustration of the principle of dark-field illumination based on Abbe theory. Sets of parallel beams fall in many different directions on to the slide and continue to a focus in the back focal plane of the objective from which they go forward to be recombined into an image. Introduction of the opaque disc to eliminate beams in the axial direction results in the absence of a zero order.

the normal microscope arrangement they can hardly be seen. Variations in thickness of transparent material do however give rise to diffraction effects and under dark field conditions the details can often be seen quite clearly. The technique is not as useful as it might at first appear to be because it is extremely difficult to interpret the images produced. For example it is impossible to distinguish the thicker from the thinner pieces of material.

The second modification is the much more useful phase-contrast system. A full description would be out of place and it will be sufficient to say that the principle is similar to that of dark field illumination except that instead of *eliminating* the central order, in phase contrast systems the *phase* of the central order is changed by 90°. In practice there are all kinds of problems connected with depth of focus and other complications which make various compromises necessary but the systems can be made to work and their great advantage is that quantitative interpretation is possible; thicker sections of transparent material can be made to appear *darker* than thinner sections.

* In order to satisfy curiosity I shall try to explain how the two systems of dark field and phase contrast work using the phasor diagram (see Section 2.4), but I must warn you that this limited explanation is not by any means completely satisfactory.

For the purposes of the explanations we imagine that we have a completely transparent object which varies very slightly in thickness and hence varies the relative phase of the light passing through it by a small amount. In an unmodified microscope it will produce a diffraction pattern which will obviously have a large central peak (because so much light is passing through the whole object) and relatively weak details further from the centre. When recombination occurs the final result must be an image with nearly constant amplitude across the field (because the object is transparent) but with a slight phase change of which we are not aware. Thus we could represent the amplitude at a particular point in the field of view by means of the phasor OP in fig. 6.21 (*a*). The phasor OP′ would represent the amplitude at a place where the object is very slightly thicker and OP″ the phase where it is slightly thinner. Since the central peak is so strong and can be regarded as a reference beam, OP, OP′ and OP″ can all be resolved into two components as shown in fig. 6.21 (*b*) in which OQ is the central peak and QP, QP′ and QP″ are in each case the contributions of all the rest of the diffraction pattern outside the central peak to the point in the object which is of interest.

We can now represent the effect of dark field by eliminating OQ. You can now see that, since OP′ is longer than QP, the thicker part will be slightly brighter: unfortunately however QP″ is the same length as QP′

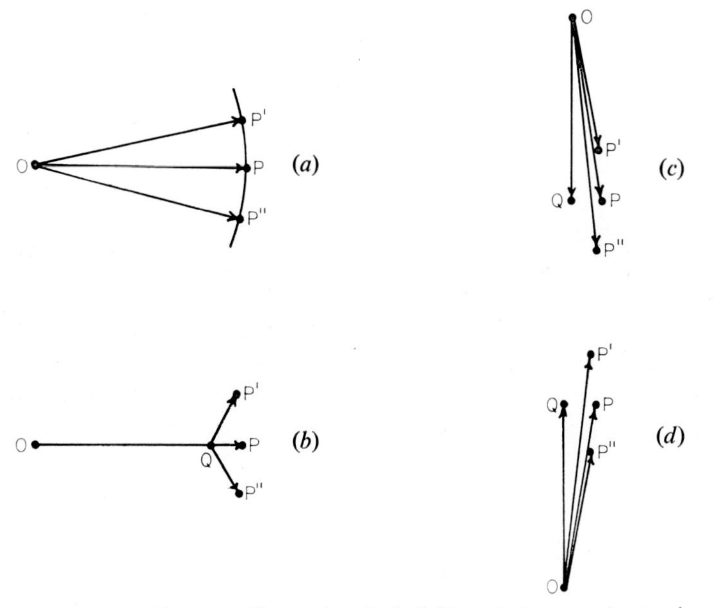

Fig. 6.21. Phasor diagrams illustrating dark field and phase contrast microscopy. (*a*) OP represents the amplitude at a particular point in the field of view: OP′ represents the amplitude where the specimen is slightly thicker and OP″ where it is slightly thinner. (*b*) Resolution of phasors into components: the effect of dark field illumination can be judged by eliminating the central order, OQ. (*c*) OQ changed in phase by 90° to give positive phase contrast. (*d*) OQ changed in phase by 90° in the opposite direction to give negative phase contrast.

and so the slightly thinner part is also slightly brighter—hence the defects of dark field as mentioned above.

The effect of phase contrast is obtained by turning OQ through 90° (fig. 6.21 (*c*)). Now OP′ is shorter than OP so the thick part appears darker; OP″ is longer than OP and the thinner part appears brighter. This is known as positive phase contrast. If the phase shift of 90° had been in the other direction the contrast is reversed as can be seen from fig. 6.21 (*d*) which illustrates negative phase contrast.

Fig. 6.22 gives illustrations of the effect achieved under phase-contrast conditions.

Now I shall turn to my historical interlude. In the 1870s microscopists used diatoms in a great deal of their experimental work because they had extremely regular features (see for example fig. 5.8) and were of such a size that many of the details of their structures were just about on the limit of optical resolution: they are still often used as test objects to assess the perfection of microscope objectives. One well known diatom used at that time, and still today, is *Pleurosigma angulatum*. It has an essentially hexagonal structure and was known to give orders of diffraction in the back focal plane of the objective (Abbe theory) as shown in fig. 6.23 (*a*).

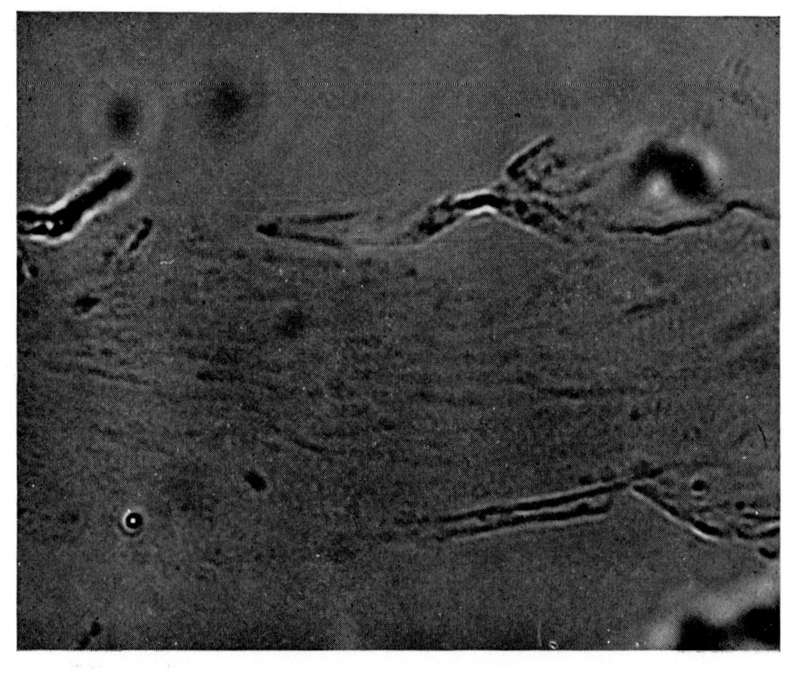

(*a*)
Fig. 6.22 (*cont. opposite*).

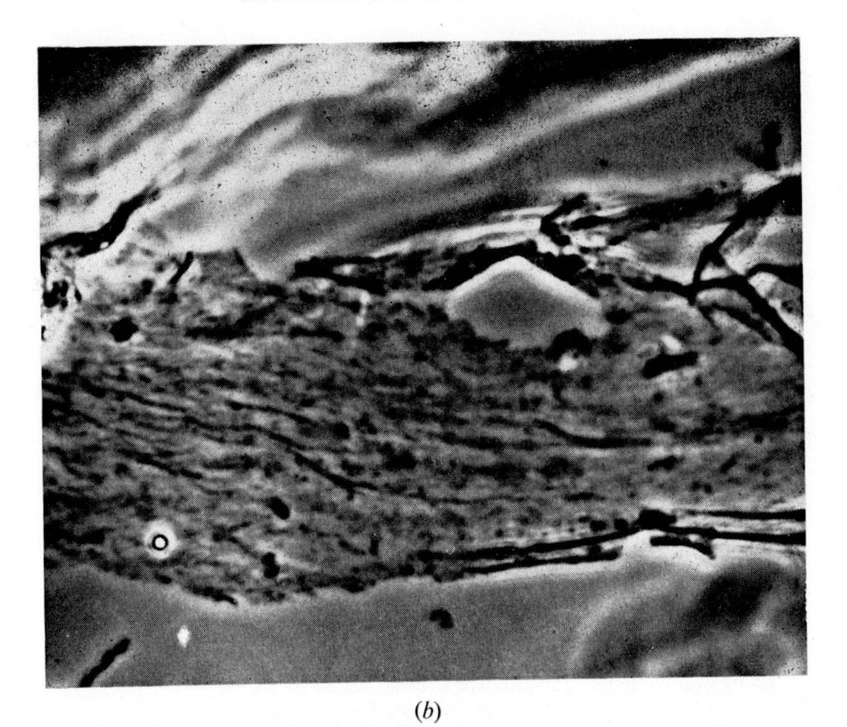

(*b*)

Fig. 6.22. (*a*) Normal micrograph (×1000) of a cotton hair that has been immersed in a solvent which removes the cellulose component and leaves the cuticle or ' skin '. (*b*) Phase-contrast micrograph (×1000) of the same specimen as in (*a*). The slight change in shape arises because between the two photographs being taken a little more of the core has been dissolved; nevertheless the enhancement of the contrast is very clear. By permission of Mr. S. Simmens, Shirley Institute, Manchester.

A Mr. Stephenson records in the *Journal of the Royal Microscopical Society* for 1878 the following incident. A mathematical student who had *never seen a diatom*, taking the spectra alone, recorded in the back focal plane of the microscope objective, (fig. 6.23 (*a*)) worked out by calculation the drawing, reproduced as fig. 6.23 (*b*), as the object that gave rise to such spectra. Fig. 6.23 (*b*) is thus his mathematically ' focused ' image of the diatom—an extremely early example of the process of indirect recombination!

Mr. Stephenson goes on to say that the *small* markings between the hexagons had never been seen in *P. angulatum* by anybody. But on his making extra careful investigations and stopping out the central pencil " so that its superior illuminating effects might not over-power the others." (i.e. he used dark field illumination) " these small markings

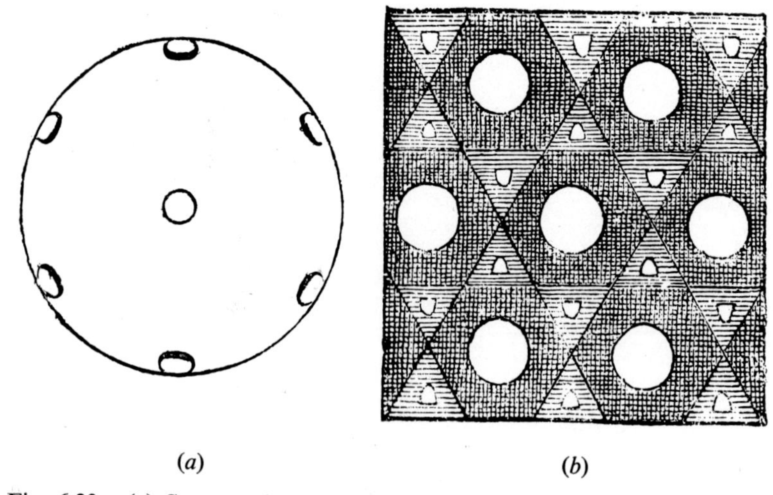

Fig. 6.23. (*a*) Spectra observed in the back focal plane of a microscope objective when a diatom is placed on the specimen table. (*b*) Calculated image deduced by a student in the 1870's. From *Light*, by Lewis Wright (1882).

were found to exist, though they were so faint as to have eluded all observations until mathematical calculation from their spectra had shown that they *must* be there." Lewis Wright reporting this in the 1890s comments " Light was once more, even in the microscope, by its physical deportment, a Revealer of what the microscope had, up to that date, failed to see."

6.9. *Electron microscopy*

In Section 3.5 we discussed lenses for electrons and indicated some of the practical problems met in trying to do with electrons what the optical microscope does with photons. The most familiar and well established kind of electron microscope is the one known as the transmission electron microscope. It can only be used with specimens up to about 5×10^{-7} m thick and the information imaged is the differential absorption of electrons by different thicknesses of material and by variations in the nature of the specimen from point to point. Although there are practical complications (resulting from the need for the whole system, including the specimen, to be in a vacuum, for the high voltage to be very precisely stabilized, for the electrons to be accelerated to sufficiently high energies to penetrate the specimen, and from the fact that the lenses are precision-made electromagnetic systems), the optical arrangement is identical with that of an optical microscope. A source of illumination (the electron gun) followed by a condenser lens (electromagnetic) irradiates the specimen with electrons. An objective lens (electromagnetic) then produces a real and highly magnified image and a projection lens (also electromagnetic) produces another real image with further magnification on a fluorescent screen which is also inside the vacuum. For record purposes a photographic plate can be inserted through a vacuum trap to take the place of the fluorescent screen.

As is the case with most sophisticated instruments, there is a great art in getting the best out of electron microscopes and the mere record of the best resolution obtained does not really give an adequate impression of the results achieved. To my mind some of the most beautiful and revealing photographs ever produced are those by R. G. Wyckoff of virus crystals. They are now quite old and have been surpassed in terms of fineness of resolution but remain in a class by themselves as a combination of scientific significance and artistry. Fig. 6.24 is a typical example.

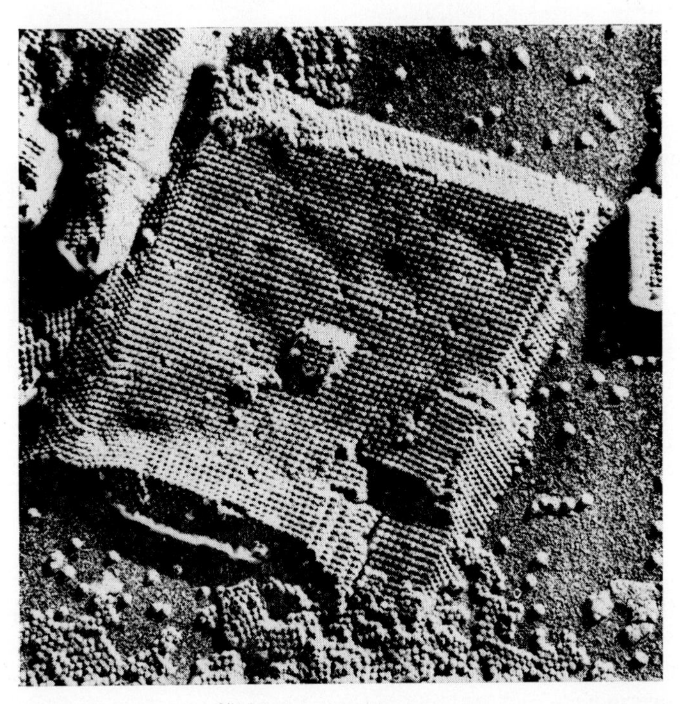

Fig. 6.24. Electron micrograph of a shadowed replica of a tobacco necrosis virus crystal. Magnification approximately 50 000. By permission of Professor R. W. G. Wyckoff, University of Arizona.

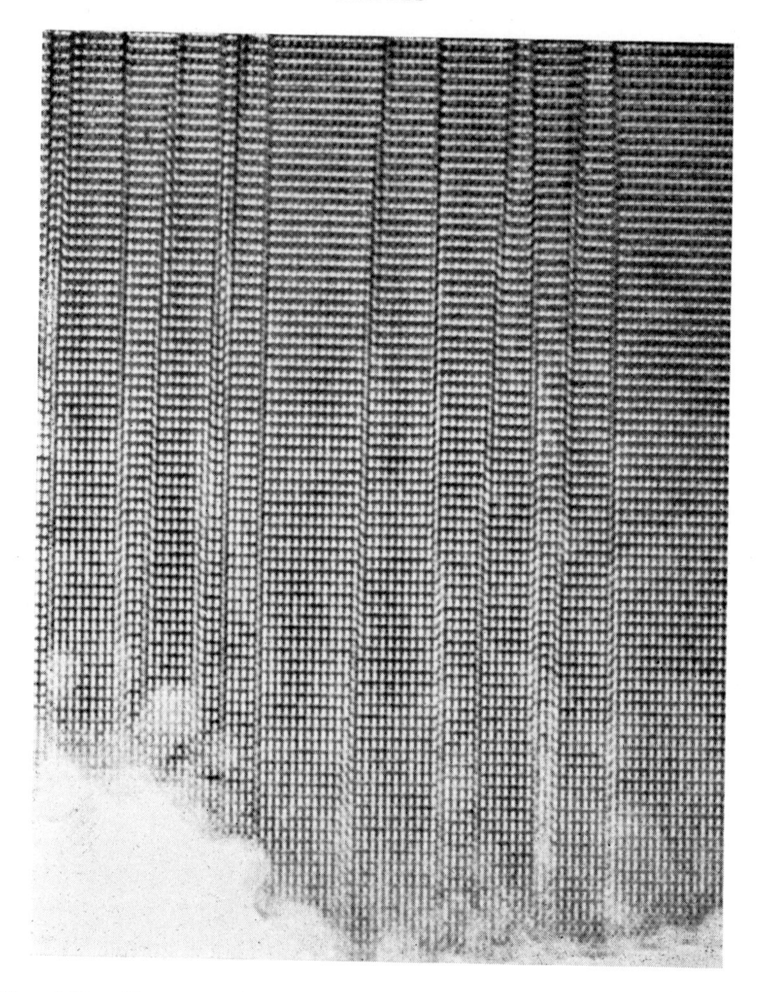

Fig. 6.25. Electron micrograph of a thin crystal of magnesium fluorogermanate. The main periodicity of the unit cells in the horizontal direction is 0·59 nm and faults in the perfection of the crystal can also be clearly seen. The magnification is approximately 3 million. By permission of Sumio Iigima, Department of Physics, Arizona State University.

One very recent example of extremely high resolution electron microscopy published in 1976, which achieves a magnification of about 3 million times, is shown in fig. 6.25. The material is magnesium fluorogermanate and each white square represents a feature of a single unit cell of the structure; they are spaced $5{\cdot}9 \times 10^{-10}$ m apart in the horizontal direction. The particular fascination is that here we can see very clearly that the crystal is not perfect, and there are stacking faults occurring at anything between 3 and 18 cells apart.

Surfaces of materials are of great interest to scientists but you will have realized that the normal transmission microscope is unlikely to be able to deal with specimens such as surfaces of fractured metal unless the remaining metal supporting the surface can be cut or ground down to a thickness of the order of 5×10^{-7} m (about the wavelength of visible light by coincidence). Some remarkably elegant ways of surmounting this difficulty have been achieved over the years. For example, it is possible to make a replica of the surface being studied by an exactly similar process to that used by a sculptor in reproducing a bronze statue from a clay or plaster mould. A liquid synthetic resin mixed with a hardener is painted on to the specimen and, in due course, when it has solidified, the replica in very thin plastic is peeled off. Now we have a specimen thin enough not to absorb too many electrons. Unfortunately, of course, it does not now absorb many at all and so it is not able to impress any particular information (in the form of a scattering pattern) on the incident electrons if it is placed in the electron microscope.

A second technique known as 'shadowing' is then brought into play. The replica is placed in a vacuum and a thin layer of metal is deposited on it by the so-called 'evaporation' process. If the metal is raised to the melting point, atoms of the vapour will escape from the surface (just as water molecules evaporate from a water surface well below its boiling point) and, because it is in a vacuum, there will be no collisions with gas molecules and the atoms will travel in straight lines. By a suitable geometrical arrangement (fig. 6.26), the metal atoms can be made to fall at an angle on the replica (R) and hence produce 'shadows' in metal. The ups and downs of the surface are thus converted into opaque and transparent metal areas and the shadowed replica now has the characteristics needed for producing good electron scattering

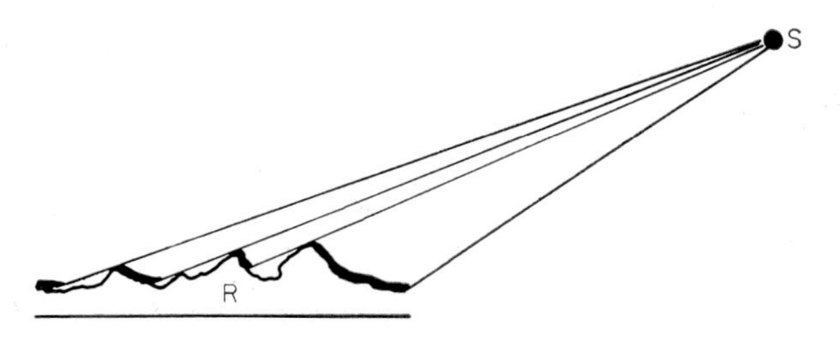

Fig. 6.26. Principle of the metal shadowing technique.

patterns and then recombined images. Fig. 6.27 shows an electron micrograph of a metal surface made in exactly this way.

But in spite of all the ingenuity in techniques and in spite of the apparently favourable wavelength of electrons (50 kV electrons have a wavelength of about 5×10^{-12} m) instrumental errors still prevent the electron microscope from achieving its full promise.

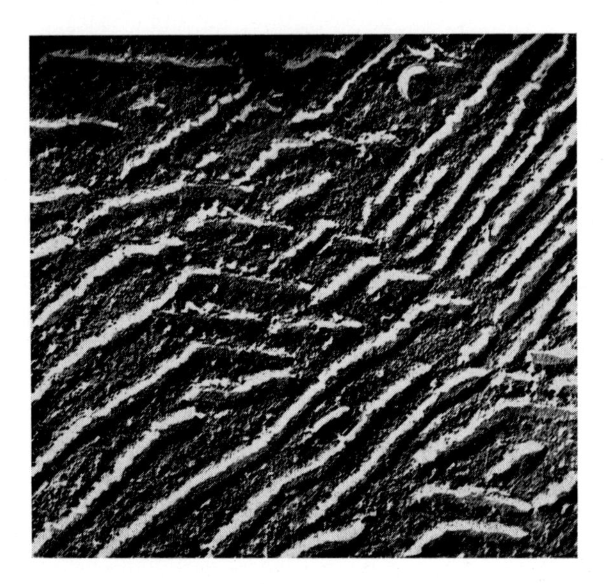

Fig. 6.27. Shadowed replica of the surface of a specimen of carbon steel: the protruding ridges are of cementite and the magnification is about 10 000. By permission of Dr. John Taylor, Department of Metallurgy, University College, Cardiff.

Lens imperfections are one of the principle defects and these arise from the fact that the precision with which metal components of the electromagnetic lenses need to be made greatly exceeds the optical perfection with which glass lenses need to be polished; so far the practical problems of achieving this accuracy for the particular shapes involved in the electron microscope lenses have not been solved.

In addition it is very easy for an electron microscope to slip out of adjustment. For example the focus might not be quite right or a deposit of tungsten evaporated from the filament may fall asymetrically on to one of the tiny aperture stops which restrict the electron beam and may produce very strange astigmatic effects. In principle what happens is that the source of electrons is no longer effectively a point source but rather acquires some odd shape. The final image is a convolution of the true image with the shape of the source and this can differ seriously from the true image alone. Further, *unless* you know exactly what the object *should* look like, you may be unaware that defects exist. Figs. 6.28 (*a*), (*c*) and (*e*) show three photographs of the same object with three different focusing adjustments and each made with a non-astigmatic system. Clearly the pictures are *different*, but without the comparison who is to tell whether either on its own is right or wrong? Fortunately, we can use a technique based on imaging and diffraction theory to test each picture.

We have already seen that the image produced is the convolution of the true image with a point spread function (see Section 5.5), which is an image of the electron source which may well be distorted. Suppose now we place the electron micrograph itself as a diffracting object in the optical diffractometer. The result will be the product of the diffraction pattern of the true image and the diffraction pattern of the shape of the image of the source. Now this source image is much *smaller* than the details of the image in which we are interested: it follows that the diffraction pattern on the spot is *much larger* than that of the details of the image. Thus the dominant feature of the diffraction pattern of the electron micrograph is the diffraction pattern of the spot shape and this can easily be interpreted. If the spot is circular and tiny the diffraction pattern is large and Airy-disc like. Figs. 6.28 (*b*), (*d*) and (*f*) show the diffraction patterns of 6.28 (*a*), (*c*) and (*e*) respectively. It is not difficult to see that the focus is different and fig. 6.28 (*g*) and (*h*) show the effect when astigmatism is present.

This sort of ' image analysis ' is becoming increasingly popular and is very helpful in preventing total misinterpretation of the electron micrographs.

The next problem in trying to push forward the limits of what can be done with the electron microscope is that it is not easy to distinguish in a transmission micrograph between changes in contrast (black, white and grey) that come from changes in specimen thickness and those that come from changes in chemical nature. A grain of the same material on the surface providing extra thickness might for example look the same as an impurity inside the crystal of the same size and shape. The system known as the microprobe analyser goes some way to resolving the difficulty. As the beam of electrons hits any particular point of the specimen characteristic X-rays are emitted and their wavelength gives information about the material producing them. By first producing a normal electron micrograph and then irradiating areas of the specimen about which further information is needed with

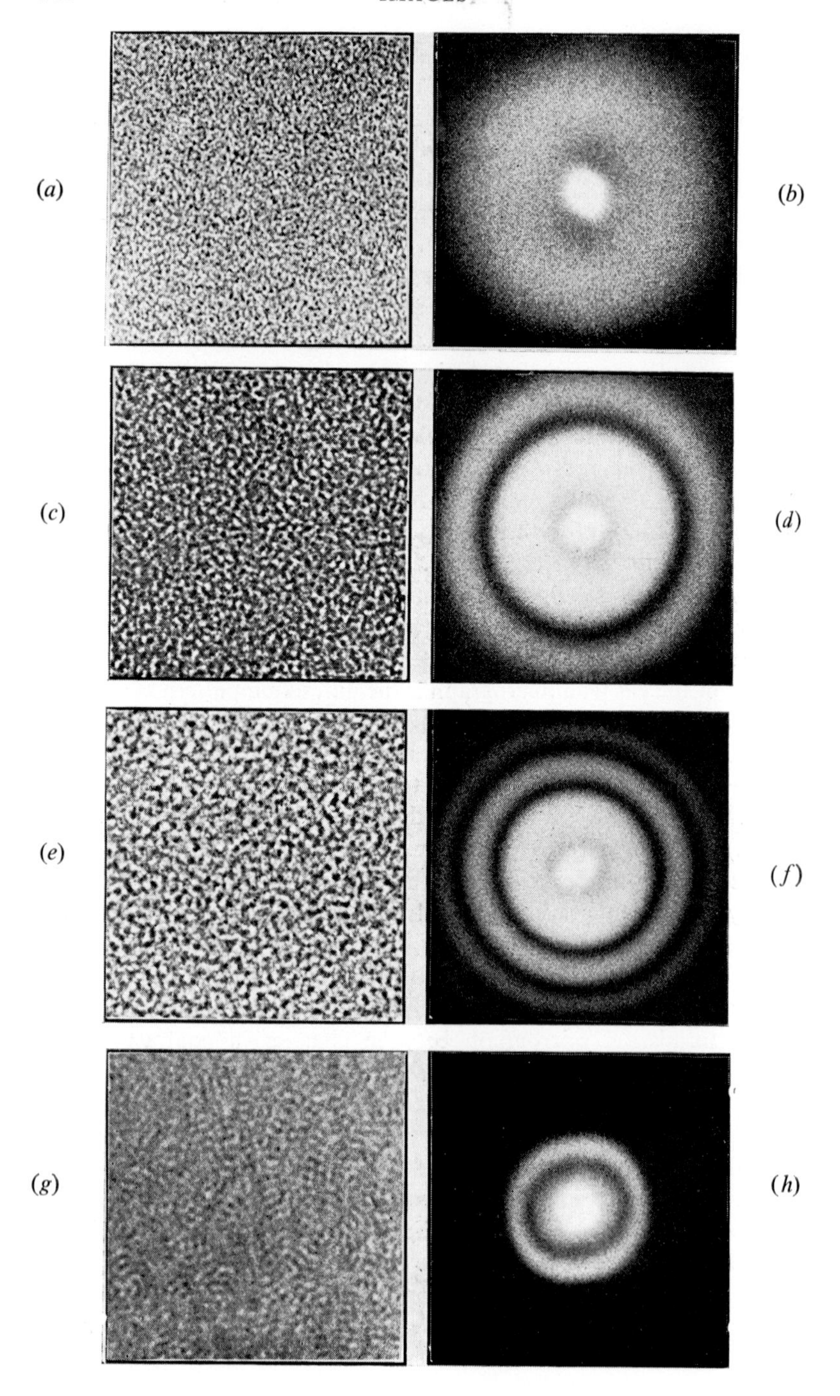

(a)
(b)
(c)
(d)
(e)
(f)
(g)
(h)

electrons and analysing the resultant X-rays, great strides in interpretation have been made.

Finally we must talk of the problem of depth of focus introduced in Section 5.5. In that section we saw that scanning the object with a narrow pencil of electrons gets over the depth problem. But what actually happens when the electrons strike the object; are they merely scattered? The answer is that they are not and indeed the charges that are collected and which form the signal are in fact *secondary* electrons that are emitted from the body of the material. It becomes clear then that the specimen is really acting as a self-luminous source, but, of course, only the spot being irradiated at any one moment is emitting the secondary electrons.

Careful investigation shows that the angle of the surface affects both the penetration of the primary electrons and also the way in which secondary electrons can diffuse out, and for this and various other reasons we find excellent representations of surfaces produced by the scanning electron microscope. The resolution of scanning microscopes has progressively been increased. Among other techniques the actual size of the primary source of electrons has been reduced by using a fine tungsten wire with a tip of radius 5×10^{-8} m as a field emission source (as for the field emission microscope). High energy electrons can be used—though of course there comes a point at which the law of diminishing returns comes in, because higher energy electrons create more damage to the specimen and so the thing we are studying may be irretrievably damaged before the image can be examined. Magnifications of up to a million in the scanning microscope are, however, quite possible and the probability of ‘ seeing ’ individual atoms is just around the corner.

6.10. *X-ray, electron and neutron diffraction*

In Chapter 4 we saw that the whole range of techniques of investigation that are grouped under the general heading of X-ray diffraction or

Opposite:

Fig. 6.28. (*a*), (*c*) and (*e*) Electron micrographs of amorphous carbon film with a magnification of about $1\frac{1}{2}$ million and three different adjustments of focus. (*b*), (*d*) and (*f*) The corresponding optical transforms of (*a*), (*c*) and (*e*). (*g*) Electron micrograph of amorphous carbon film at a magnification of about 5 million but with an astigmatic condition. (*h*) Optical transform of (*g*): the oval shape of the central peak reveals the astigmatism. Photographs (*a*), (*c*) and (*e*) by Miss D. Chescoe, A.E.I. Harlow; (*b*), (*d*), (*f*) and (*h*) by Dr. D. Somerford, University College, Cardiff; (*g*) by F. Sheldon, Pye-Unicam, Cambridge.

X-ray crystallography can be quite properly considered as aspects of image formation. With electrons, it is perfectly possible, as we saw in the last section, to focus images; but there are some occasions when, for a variety of reasons, it can be more useful to interpret electron diffraction patterns directly instead of allowing the recombination to occur in the electron microscope. In particular this is true when the detail being studied is of atomic size and the aberrations of the image-recombination system introduce resolution limits that are much coarser than the theoretical ones for electron wavelengths. The basic principles of interpretation of electron diffraction patterns are broadly similar to those for X-ray diffraction, though there are inevitably some differences in detail. It is interesting in passing to note that electron diffraction was not discovered until about 15 years after the discovery of X-ray diffraction.

Neutrons may also be used but here, because they are uncharged, there is no way of recombining them directly and interpretation of the diffraction pattern is the only avenue open. The special advantage of neutrons is that—again because they are uncharged—they are not scattered by the electron cloud round an atom but are scattered by atomic nuclei. The scattering effects are similar for all atoms and do not, as in the case of X-rays, depend on the atomic number. Thus hydrogen atoms, which are difficult to locate by X-ray diffraction techniques, may be located relatively easily by neutron diffraction.

In our discussions of X-ray diffraction in Chapter 4 we concentrated on illustrating the technique of direct interpretation of the scattering pattern without recombination. But, of course, one alternative approach is not to bother about interpretation at all, but merely to regard the scattering pattern as a kind of 'finger-print' which can be used for identification without interpretation. This is the basis of the so-called 'powder' method that has had enormously successful applications over the years. A high proportion of solid materials, even when existing as finely divided powder, has a very pronounced regularity of structure and consists essentially of collections of tiny crystallites each of which is really a complete crystal covering many unit cells. Each of these tiny crystallites would give an X-ray diffraction pattern consisting of a set of sharp spots on a regular array (such as fig. 4.2 (*a*)). Since a powder consists of many thousands of such crystallites in all possible orientations, each crystallite will give the same pattern, but patterns will occur in all possible orientations. The resulting diffraction pattern will thus consist of sets of concentric rings whose diameters and intensities will be characteristic of the material. Fig. 6.29 (*a*) shows such a photograph of a material in which the crystallites are relatively large and not too numerous; the separate spots building up the rings can still be seen. Only short segments of the rings are recorded. The material here is coarsely crystallized corundum—Al_2O_3—such as is used for grinding lenses. In 6.29 (*b*) the material is the same but the crystallite size is much smaller and the spots fuse together into smooth rings. In 6.29 (*c*) the more usual form of photograph on a strip of film is shown; here the material is the mineral rutile (titanium dioxide, TiO_2).

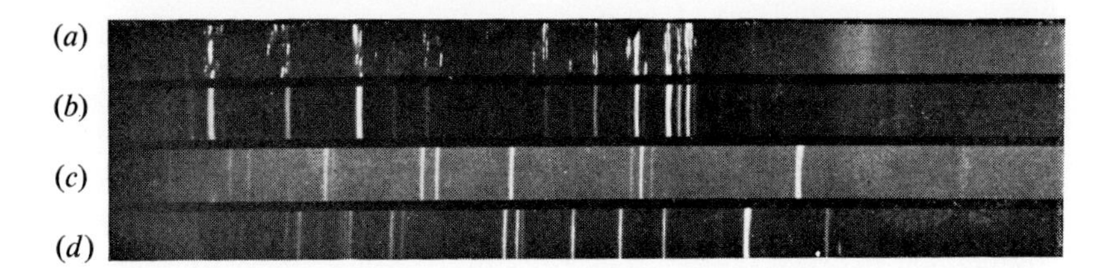

Fig. 6.29. Four ' powder ' diffraction photographs. (*a*) Coarse sample of corundum (Al_2O_3). (*b*) Very finely divided sample of corundum. (*c*) and (*d*) Two chemically indistinguishable samples of titanium dioxide. The crystal structures are quite different as is revealed by the diffraction patterns. (*c*) is rutile and (*d*) is anatase.

It is important to realize that each photograph of this type is not characteristic of a particular *chemical* structure but of the particular crystalline or structure variant of the material too. Fig. 6.29 (*d*) is also of titanium dioxide, TiO_2 and the specimen is *chemically* identical with that for 6.29 (*c*), but here the crystalline form is that for the mineral anatase and the totally different powder diffraction pattern is quite obvious.

One example of purely routine use which immediately jumps to my mind relates to the glass industry. In Section 6.6 we mentioned that certain kinds of glass, if raised to too high a temperature, attack the lining of a furnace or ladle. If a piece of solid material is found in a batch of glass it is clearly important that the source should be tracked down as soon as possible. An X-ray photograph of the inclusion can be taken in a matter of a few minutes and the nature of the material will indicate its source; if there are several ladles or furnaces with chemically identical linings it is even possible to include in their manufacture different chemical markers which can be immediately identified from powder photographs.

X-ray diffraction does not occur only with highly crystalline materials. Indeed some of the most challenging patterns to the interpreter come from materials that are not crystalline at all in the simple sense. The vast array of new materials classified under the general terms ' plastics ' or ' polymers ' are usually not highly regular crystals but the individual long chain molecules which characterize them do have regularity of a kind and this imposes itself on their X-ray diffraction patterns. Fig. 6.30 shows two examples from well known materials and it is clear that they neither have the sharp spots on a lattice characteristic of single crystals nor the rings characteristic of aggregates of tiny crystallites. Nevertheless, by the study of the distribution of the blackening of the film, a surprisingly large amount of information about the nature and structure of the material can be deduced if the research worker is familiar with the relationships between objects and their diffraction patterns.

To complete this selection of X-ray diffraction illustrations, I propose to give, in somewhat greater detail than in Chapter 4, an example of a comparison between optical and X-ray diffraction patterns for a particular material; I hope this may prove valuable as a means of consolidating firmly the ideas about relationships between objects and their diffraction patterns that were established in the main text.

Fig. 4.11 illustrated the final answer to the solution of the structure of hexamethylbenzene deduced many years ago by the classical methods of X-ray crystallography. Now we will look at some of the steps that might have been taken if this structure had been studied by optical analogue methods.

Fig. 6.31 (*a*) shows an X-ray photograph taken by the precession method (Section 4.1). Fig. 6.31 (*b*) shows an idealized diagram of one molecule in which all twelve carbon–carbon bonds are assumed to be the same length. The inner six carbon atoms form the benzene ring and the outer six are the nuclei of the six methyl groups. The hydrogen atoms are ignored for our present purposes.

This particular structure has been chosen because it has only one molecule in each unit cell and this greatly simplifies the geometrical problems and the argument—but I must hasten to tell you that such examples are rare.

The X-ray picture of fig. 6.31 (*a*) enables us (knowing the dimensions of our apparatus and the wavelength of the X-rays) to deduce immediately two of the dimensions of the unit cell of the crystal and, if we assume the crystal to be a convolution of one molecule with the lattice, the photograph should be the *product* of the diffraction pattern of the lattice with the diffraction pattern of one molecule. The diffraction pattern of the lattice is often called by crystallographers the ' reciprocal lattice '.

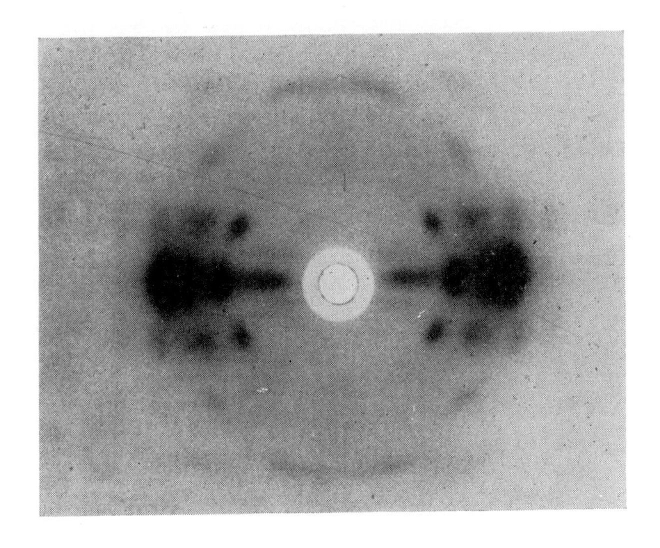

(*a*)

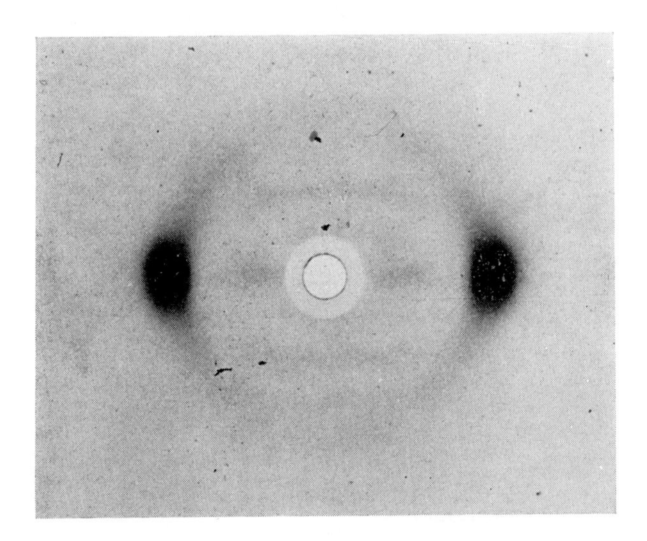

(*b*)

Fig. 6.30. X-ray diffraction patterns for (*a*) a polyester fibre (e.g. ' Terylene '), (*b*) a polyamide fibre (e.g. ' Nylon ').

Fig. 6.31 (*c*) is the diffraction pattern of the molecule as shown in fig. 6.31 (*b*) and we need to compare it with fig. 6.31 (*a*). Can you see six groups of spots near the edge of fig. 6.31 (*a*) that are *not* arranged as a regular hexagon but as an elongated hexagon? To help identification their position is indicated in fig. 6.31 (*d*). These could correspond to the six circular peaks near the edge of 6.31 (*c*) if the molecule, were distorted in some way. Suppose we *tilt* the molecule so that, in projection, it looks like 6.32 (*a*). Now we have distorted

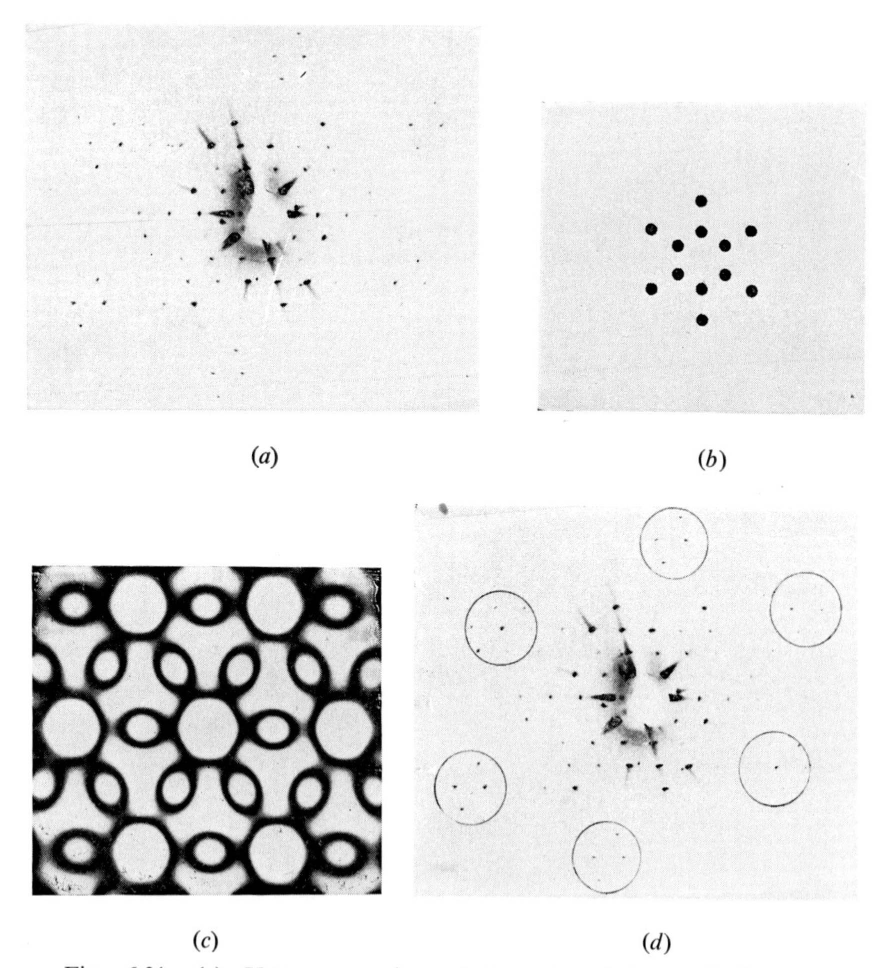

Fig. 6.31. (*a*) X-ray precession photograph of hexamethylbenzene. (*b*) Diagram of one molecule of hexamethylbenzene. (*c*) Diffraction pattern of (*b*). (*d*) Location of groups of spots in (*a*) corresponding to the six peaks of (*c*).

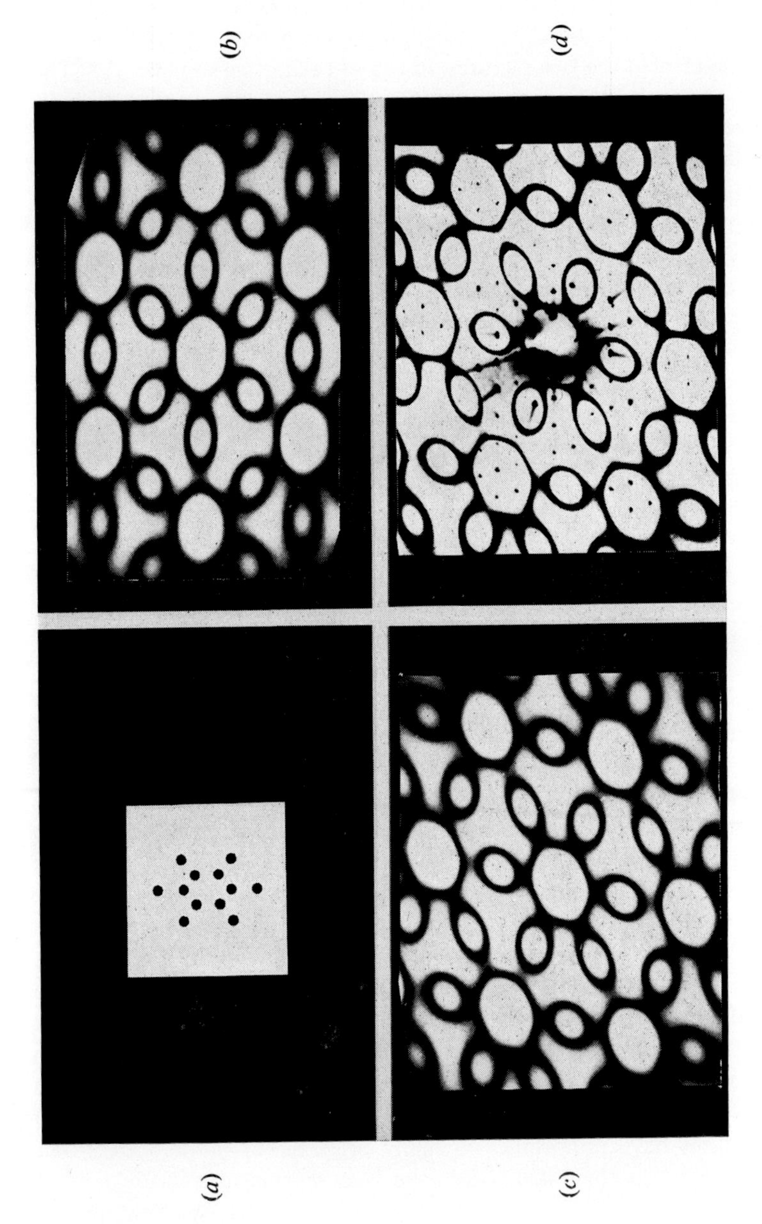

Fig. 6.32. (*a*) Molecule of 6.31 (*b*) tilted. (*b*) Diffraction pattern of (*a*). (*c*) (*b*) rotated to match 6.31 (*d*). (*d*) (*c*) superimposed on 6.31 (*a*).

the molecule and its diffraction pattern (6.32 (*b*)) resembles the arrangement of strong spots in 6.31 (*d*) provided that we also rotate it. Fig. 6.32 (*c*) shows it rotated into the right position to match 6.31 (*a*) and we have printed the two pictures on top of each other in 6.32 (*d*). The outer six peaks match reasonably well and so we appear to have deduced the orientation of the molecule in space.

This is by no means the whole answer and, as already said, this is a particularly easy example; I hope, however, that it gives some idea of the kind of deductions that it is possible to make. It should certainly help you to see how some of the relationships between objects and their diffraction patterns—the Fourier transform relationship—work out and become of great practical value.

Now let us turn to electron diffraction and consider first how the patterns are actually observed. Various special arrangements have been devised from time to time but by far the simplest technique is exactly parallel with that used by Abbe in his studies of the optical microscope. The distribution of intensity in the back focal plane of the objective of an optical microscope can be studied by rearranging the microscope so that the eyepiece focuses on this plane instead of on the conjugate image plane of the object. With an electron microscope the projection lens which forms the image is re-adjusted so that it images the back focal plane of the objective and the electron diffraction pattern then replaces the image on the screen or photographic plate. One of the study areas in which this technique has proved particularly informative is that of thin film technology. We mentioned vacuum evaporation as a means of shadowing replicas of surfaces for use in the electron microscope in the last section. If, in the evaporation chamber, two or more metals are evaporated simultaneously and allowed to form a thin film on a base, or substrate as it is usually called, (it is often a polished rock-salt, quartz or other crystal surface) then the resulting film is a very special kind of alloy. It is special because it has been built up in such a way and the atoms of the various metals present have a degree of ‘ choice ’ in how they arrange themselves in the layers. In the more usual processes of crystallization from molten metals the build up is much more rapid and, of course, grows outwards from nucleation centres in the solid instead of in successive two-dimensional layers as in evaporation. The electrical and magnetic properties of these films have importance in many branches of modern electronic technology, but their structures have been worked out in great detail from electron diffraction studies. Fig. 6.33 shows two examples of electron diffraction patterns which are both beautiful because of their high symmetry and also very full of structural information. It is important to notice that for an electron diffraction pattern produced by readjusting a typical microscope, with an accelerating voltage of say 50 000 V, the electrons have a de Broglie wavelength of about 5×10^{-12} m. This is about 1/100 of typical atom–atom spacings in solid matter and this kind of wavelength ratio compares with the wavelength-to-object size ratio commonly occurring in optical diffraction. Our electron diffraction pattern is therefore of relatively small angular extent and is very closely parallel in geometry to the optical diffraction pattern produced in the diffractometer illustrated in fig. 2.20.

Although there are many other applications of electron diffraction that are important I just want to include one further development. A very significant group of researches uses electrons of very small energy. The field is usually known by the acronym LEED (Low Energy Electron Diffraction). The accelerating voltages used in practice are in the region 10–500 V and the corresponding de Broglie wavelength is 4×10^{-10} to 5×10^{-11} m. This is in the same size range

(*a*)

Fig. 6.33 (*cont. overleaf*).

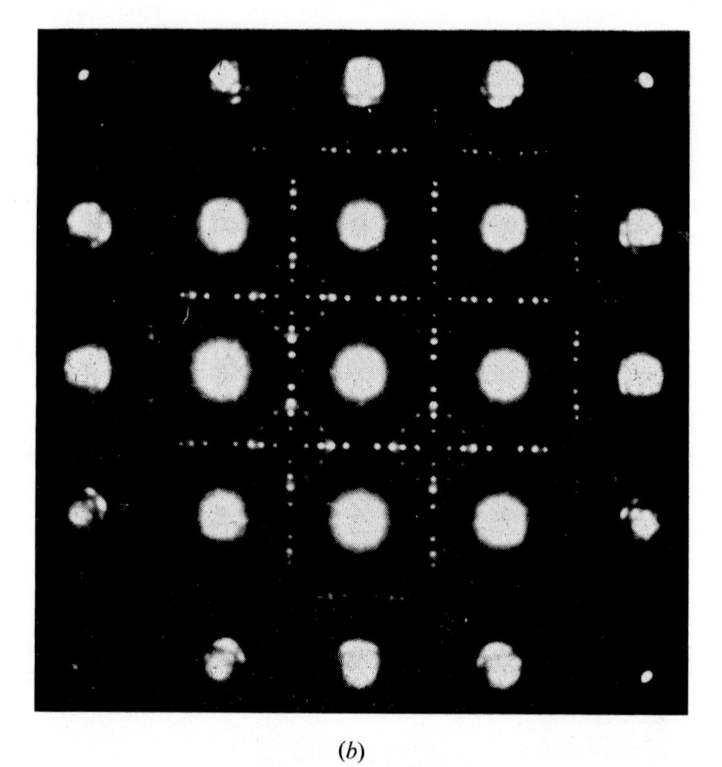

(*b*)

Fig. 6.33. (*a*) Electron diffraction pattern of a thin film of an alloy of gold and zinc. (*b*) Electron diffraction pattern of a thin film of an alloy of gold and manganese. By permission of Dr. W. S. Michael, Department of Physics and Mathematics, Manchester Polytechnic.

as atom–atom spacings and hence the scattering geometry is likely to be much more like that of X-ray diffraction where diffraction through angles all the way up to 180° occurs. The experimental arrangement for LEED is therefore quite different from that for conventional electron diffraction. The schematic arrangement is shown in fig. 6.34. The gun irradiates the specimen and the scattered and secondary electrons pass back through a system of grids which permit them to be accelerated sufficiently to create luminous spots on the phosphor on the screen without changing their direction of travel. With such low energies it is the surface structure that is revealed in the luminous pattern on the screen. Gas and other impurity atoms on the surface presented great difficulties in the early days but very high vacuum systems and special cleaning techniques have been developed to such a stage that very useful surface information may now be obtained. Surface

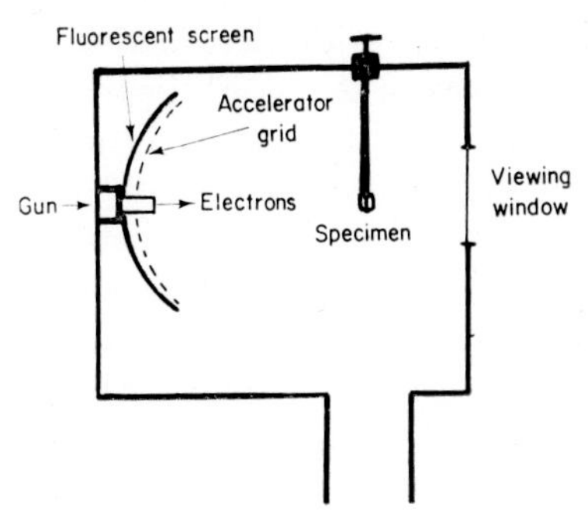

Fig. 6.34. Schematic diagram of a Low Energy Electron Diffraction (LEED) system.

behaviour plays an important role in many modern technologies—catalysis, corrosion, solid state devices etc., and LEED has already fed in a great deal of valuable information for workers in these and other fields.

The wavelengths corresponding to neutrons of thermal energies are much the same as those of the X-ray range. For example, at 0°C the neutron wavelength would be $1{\cdot}55 \times 10^{-10}$ m and at 100°C it would be $1{\cdot}33 \times 10^{-10}$ m. The common X-ray wavelength used in crystallography—the characteristic radiation from a copper target—is $1{\cdot}54 \times 10^{-10}$ m. The actual techniques used in neutron diffraction studies resemble in principle those used for X-ray diffraction. A collimated beam of neutrons emerging from a nuclear reactor falls on a crystal and the diffracted beams are detected and measured by means of an electronic counter. The interpretation of the scattered beams then follows a similar trial-and-error process to that used for X-rays. The scale of the apparatus is much larger: the collimator for example has to be very massive indeed in order to absorb all neutrons but those going in the required direction. The neutron flux from the reactor is not very high and relatively large crystals are needed to give easily detected beams. The protective shielding required on all parts of the apparatus increases the size and mass enormously and this influences to some extent the kind of materials and problems that can be studied.

Neutron diffraction is rarely used on its own. In general one starts with an X-ray diffraction study and then, having extracted as much information as possible from that a neutron study can be made to give very useful additional information which is frequently complementary to the X-ray data.

I said earlier that neutrons are scattered largely by the nucleus and not by the electrons. This is not universally true and there are in fact very important interactions with electrons in the case of magnetic materials. Indeed, perhaps the most important contribution that neutron diffraction has made to the study of solids has been from studies of magnetic materials. The detailed study of the degree of order in ferromagnetic and antiferromagnetic alloys made possible by neutron diffraction is one of our principal sources of information on the atomic mechanism of magnetic behaviour. What makes this possible is the fact that an atom with its magnetic moment directed (say) upwards behaves differently towards neutrons from one with its moment directed downwards. Thus a crystal of an alloy of gold and manganese might have manganese atoms at the corner of each cubic unit cell and a gold atom at the centre of each. X-ray diffraction studies would reveal this quite easily (compare the discussion on zinc sulphide in Section 4.1).

A neutron diffraction study however would distinguish between manganese atoms having magnetic moments in different directions, and a comparison of the two sets of results would reveal a great deal about the magnetic distribution of the manganese atoms.

Among other important contributions from neutron scattering are the detection of ' light ' atoms in the presence of much heavier ones. X-ray diffraction runs in to difficulties, for example, in detecting something like oxygen in the presence of heavy atoms like tungsten or gold. The scattering from the heavy atoms is so great that it completely masks the effect of the lighter ones. The differences in scattering for neutrons are very small and many otherwise difficult problems have been resolved by comparison of the two techniques. Work on the structure of ice, for example, in which X-rays are hardly able to reveal hydrogen positions, work on uranium hydride, potassium fluoride and many other compounds of light and heavy atoms has been possible with neutron scattering.

6.11. *Radar*

We take radar very much for granted these days, and, apart from noticing the spinning or rocking antenna on ships, hovercraft and airport control centres, and assuring ourselves that our aeroplane can fly quite happily in dense cloud because of its radar, most of us give it little thought. It fits very well in to our scheme of image producers however and deserves a few pages to itself.

It has its origins almost as far back as the discovery of radio communication. It was very soon found that the geometry of a radio antenna determined the direction from which it could receive the strongest signals. An antenna mounted on a rotating turntable could therefore give an indication of the direction from which radio signals were coming. The second world war, of course, provided enormous stimulus to developments of this technique and very soon instead of just locating the direction from which a signal came, signals were being radiated from a transmitter and the *reflections* or scattered radiation from surrounding objects—particularly aeroplanes—could be picked up and their direction determined. Once the idea of transmitting the signal outwards and detecting the scattered radiation on return had been grasped it immediately became possible to determine the *distance* of the scattering object as well as its direction; in effect all that is necessary is to time the delay between initial transmission and receipt of the returned wave. Radio-location had arrived. The American term radar was soon adopted: it is a contraction of ' RAdio Direction And Range '.

In principle there are several ways in which the essentials of a radar system can be arranged. It is almost invariable to use a very short burst or pulse of radio waves as the signal in order that the timing—needed to measure the range—can be as precise as possible; usually too, the pulses follow each other at regular intervals, say 500 per second, so that changes in range or direction can be followed. The variations are: a pulse can be ' broadcast ' in all directions and then a receiving antenna can rotate to find the direction from which the required reflection is coming; the transmitted pulse can be beamed in precisely the same direction as that in which the receiving antenna is trained and the two can be rotated or rocked simultaneously; the system may use quite long radio waves which will travel huge distances and make detection possible from far away, but inevitably the resolution is very poor; very short waves can be used which will give highly detailed images of the surroundings but are limited in range by the curvature of the Earth just as light beams would be—though of course radar beams of this kind are not obstructed by mist, cloud or fog; the display of the final result may be in a ' technical ' form which requires an expert to interpret it or it may be presented just as though it was a television picture which is much easier to understand immediately.

In spite of all these variations you can see that the basic idea conforms precisely to our image-forming principles: the object is irradiated and the required information extracted from the scattered radiation. The first point to consider then is the interaction of the incident radiation and the object, and this depends chiefly on the wavelength selected. Long-range radar might use radiation with a wavelength of tens of metres in order to avoid the so-called ' optical limitation '; that is to be able to ' see ' beyond the horizon. Our usual diffraction principles tell us, however, that such interactions will be able to give only very crude indications of the presence or absence of an object of the size of a small aeroplane, with no detail whatever. As the wavelength used becomes shorter, so the resolution increases and systems operating with wavelengths of a few millimetres can be made to give very detailed images.

The second question to consider is that of whether direct recombination is possible. There are two difficulties. It is possible to refract short radio waves—indeed there are school experiments on microwaves in which a large lens made of paraffin wax is used to focus a beam. To make such a lens sufficiently free from aberrations would be quite a challenge but, even if the lens problem could be solved, there are two further difficulties. First, the magnitude of the returned signal is usually quite small and very considerable amplification is needed before it can be satisfactorily detected and it would be a very difficult problem to do that simultaneously for all parts of the scattered wave. Second, we do not have a photographic film or a photo-sensitive surface (such as is used in a television camera) that will detect the whole of a two-dimensional pattern simultaneously. The solution adopted is to use the scanning principle. The antenna may be some kind of highly directional array (like a television aerial) for the longer wavelengths or a parabolic reflector or horn at the shorter wavelengths. Often the same antenna is used both for transmission and reception with a rapid electronic switching device that will prevent the outgoing pulse from swamping the receiving circuits. In the crudest kind of system a linear cathode-ray oscilloscope scan is triggered at the moment a given pulse is transmitted and travels horizontally from left to right. The highly amplified reflection is fed to the vertical deflection plates and the distance of the object producing it can be deduced from the known speed of radio waves and the known rate at which the cathode ray tube beam is travelling in the X direction. The whole system is synchronized so that the pulses go out at, say, 500 times per second, and for each the trace on the tube will be re-produced. The antenna can be rotated until a particular peak or ' blip ' is of maximum height and then both its direction and distance are known.

In sophisticated systems, the antenna is made to scan the field of view either by rotation or by rocking, or in some other combination of movements and the cathode ray tube display is simply arranged so that the electron beam makes movements on the screen which relate to the movements of the antenna. Fig. 6.35 (*a*) shows a schematic diagram of the cruder type and fig. 6.35 (*b*) shows a radial scan system which

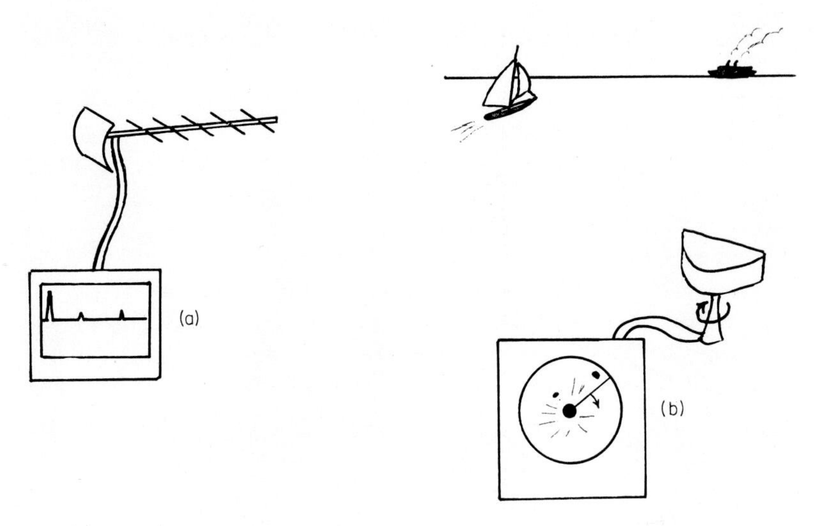

Fig. 6.35. (*a*) Diagram of radar with simple linear scan. (*b*) Diagram of radar with simple radial scan.

gives a rather more detailed picture of the surroundings. Fig. 6.36 shows a model radar system using a rotation scan and a radial trace but in this case with an ultrasonic beam instead of radio waves. It was built more or less to a design published in *Wireless World* in 1968 and gives an interesting display of objects in a lecture theatre. In the photograph an assistant is showing how separate reflections from his body and from his two hands are produced and the relative movements of his hands can also be displayed.

In addition to the obvious military and navigational uses, radar has played a considerable part in aerial surveying operations where it has two advantages over photography using visible light. The first is that mist, fog and clouds are penetrated very easily and do not hold up activities, and the second is that the reflection coefficients for different crops or land features are quite different from those of visible light or of infra-red. It is often possible therefore to identify features on large photographs taken from a very considerable height. Fig. 6.37 (*a*) shows a radar image of farmland in Kansas, in which, for example, the patches of almost white appearance are in fact fields of sugar beet. Fig. 6.37 (*b*) shows a radar image of part of the San Andreas fault in which the rock structures can clearly be seen but the dense vegetation is transparent at this particular wavelength and so does not obscure the features which geologists could not otherwise see.

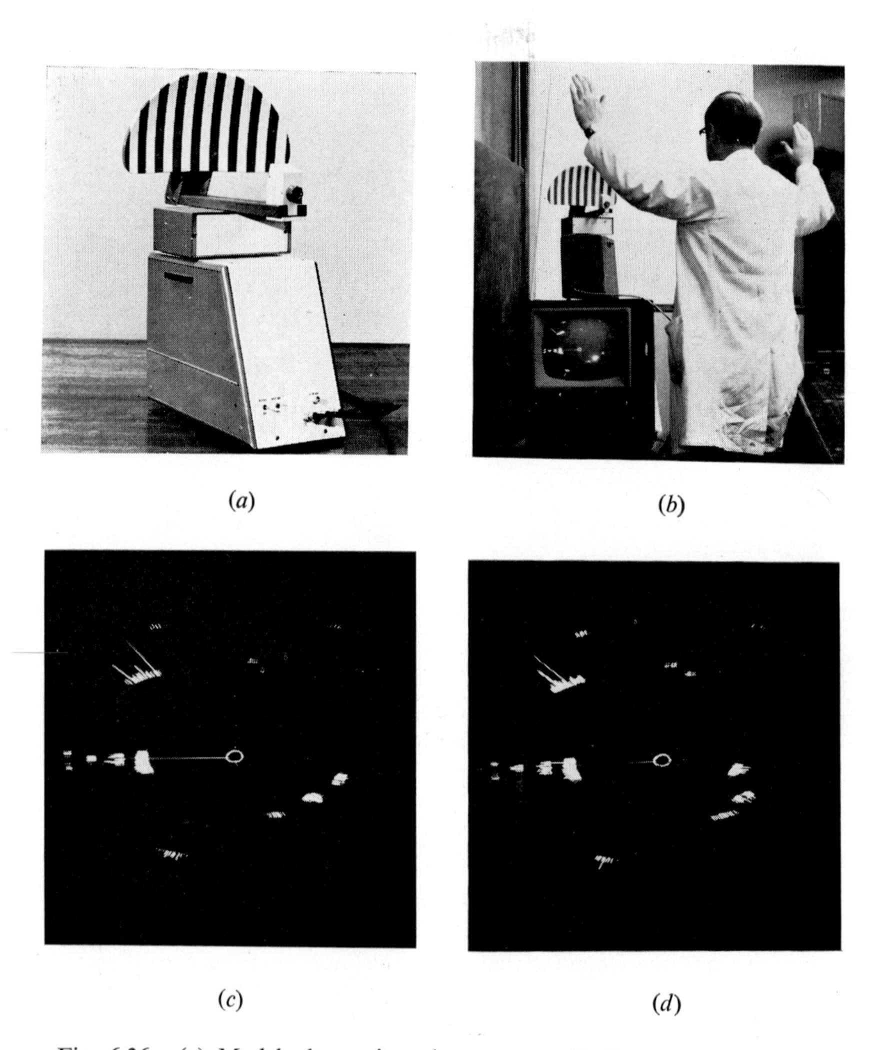

(*a*) (*b*) (*c*) (*d*)

Fig. 6.36. (*a*) Model ultrasonic radar system. (*b*) System in operation. (*c*) Scan showing location of man and hands as in (*b*). (*d*) Scan with position of hands only changed.

(*a*)

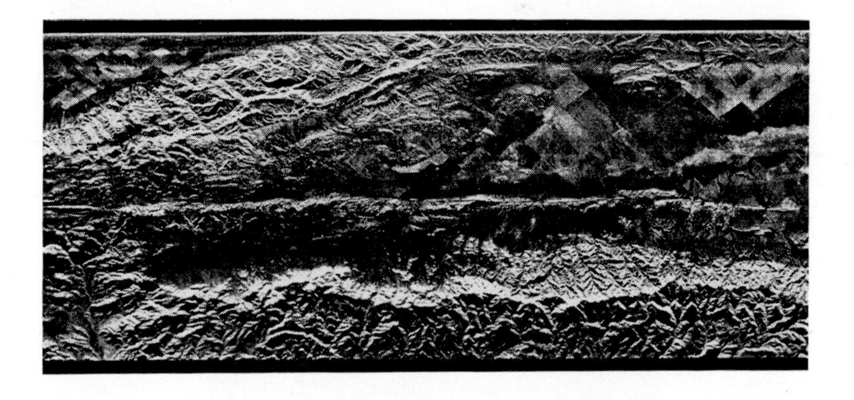

(*b*)

Fig. 6.37. (*a*) Radar image of farmland in Kansas: the almost white patches are fields of sugar beet. (*b*) Radar view of the San Andreas fault in California showing formation of rock structures underlying vegetation which does not reflect at the particular frequency used. By permission of Westinghouse Corporation, Washington, D.C.

6.12. *Medical imaging techniques, including the use of X-rays, ultrasonics, gamma rays and infrared rays*

Most people have been X-rayed at some time or other, but of course the usual radiograph taken for diagnostic purposes is sometimes not regarded as an image but as a shadow or silhouette. Nevertheless, it conforms to our usual principle of scattering and recombination with the same geometry as that of the field-emission or field-ion microscope (Section 3.2). The principal difference is that in the microscopes the object is very close to the source in order to achieve large magnifications, whereas for radiographs the object is close to the screen and produces an image practically the same size as the object (fig. 6.38 (*a*)). In order to improve the contrast and clarity, a special screen is sometimes placed between the object and the film; in effect it consists of a series of holes in a lead screen which run parallel to the direction of the X-rays from the target of the tube and which will cut out any scattered radiation which is not going in the ' right ' direction. The principle difficulty in interpreting X-ray photographs is that there is no depth discrimination; the blackening of the film depends purely and simply on, and is inversely proportional to, the total absorption between the source and the film. It is immaterial whether the object is a very thin strong absorber or a very thick weak absorber—the result depends only on the total absorption. Similarly the radiograph does not distinguish between objects at different levels (fig. 6.38 (*b*) and (*c*)).

(*a*)

Fig. 6.38 (*cont. opposite*).

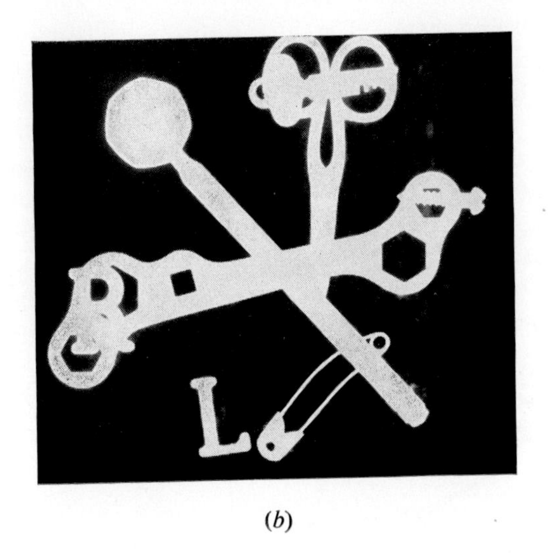

(*b*)

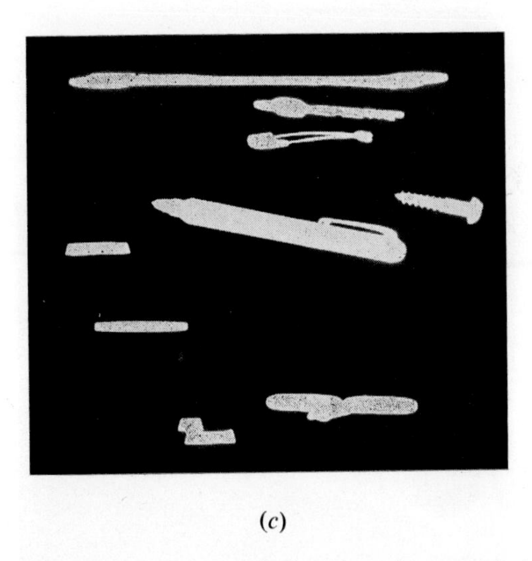

(*c*)

Fig. 6.38. (*a*) Photograph of foam block with various metal objects inserted at different levels. (*b*) Normal radiograph of (*a*). (*c*) Side-view radiograph showing different levels. By permission of Professor K. T. Evans, Welsh National School of Medicine.

In recent years a number of ingenious systems have been introduced which to some extent allow these problems to be surmounted. Two will be described here. The first is the so-called tomograph. A mechanical transport system is introduced for both X-ray tube and film. The patient remains fixed, but during an exposure the X-ray tube travels in the opposite direction to the film (fig. 6.39 (*a*)). If the speeds are carefully adjusted a line joining the source to the film can be made to rotate about a point at a given level in the patient; then for a plane through the patient at that level there will be no relative motion between the shadow and the film and hence a clear radiograph is produced. Material above and below this plane produces totally blurred images which can be ignored (fig. 6.39 (*b*), (*c*) and (*d*)).

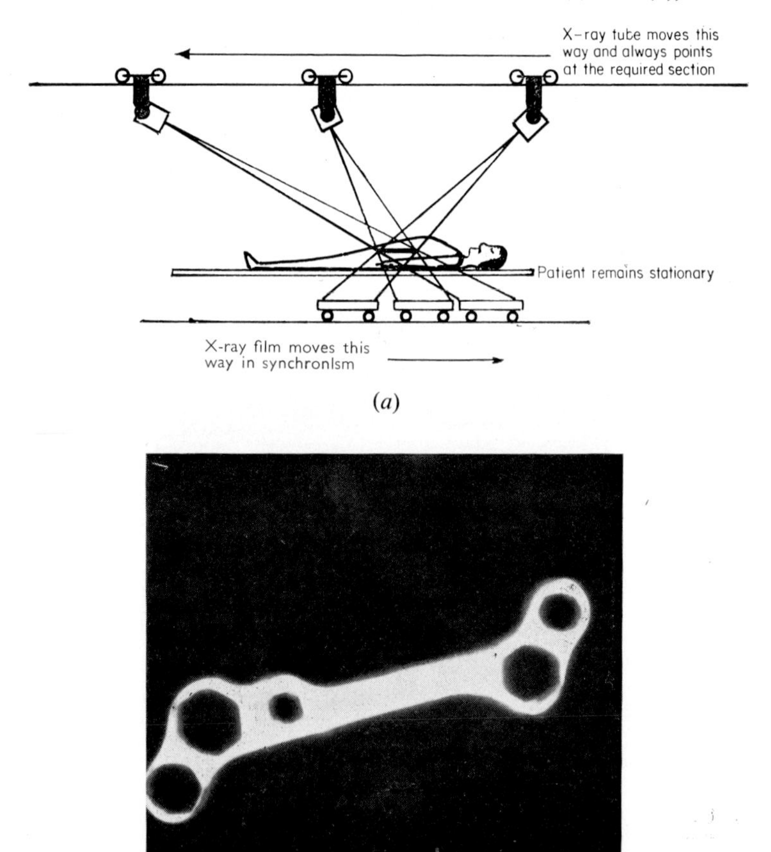

(*a*)

(*b*)

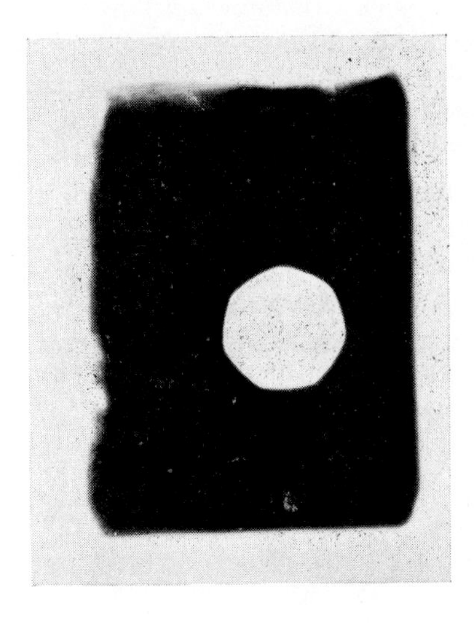

(*c*)

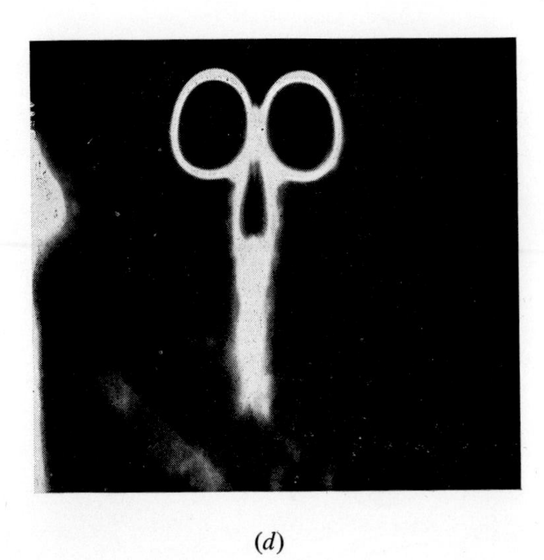

(*d*)

Fig. 6.39. (*a*) Schematic diagram of tomograph system. (*b*), (*c*) and (*d*) Tomographs picking out different objects by imaging different levels of the object of 6.38 (*a*). By permission of Professor K. T. Evans, Welsh National School of Medicine.

The second system permits the production of very detailed sectional views through quite thick structures—such as the human trunk. The technique involves the production of a very large number of standard radiographs through the object being studied, each using the X-ray beam in the plane of the required section but in a different direction (fig. 6.40 (*a*)). We thus have a great deal of information and the problem is to disentangle it. The ' focusing ' of the image is done by computer and the result is displayed on a television screen. Fig. 6.40 (*b*) and (*c*) show particularly beautiful examples. The process of unscrambling the data from the large number of views by computer has been illustrated by an analogue suggested by Professor Vainshtein of the USSR and a simplified version is presented here.

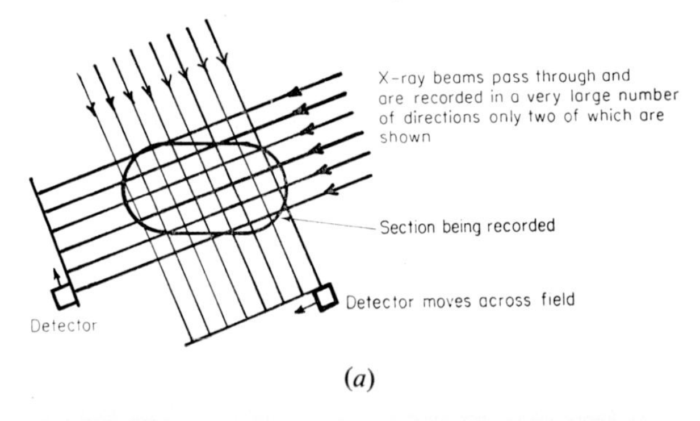

(*a*)

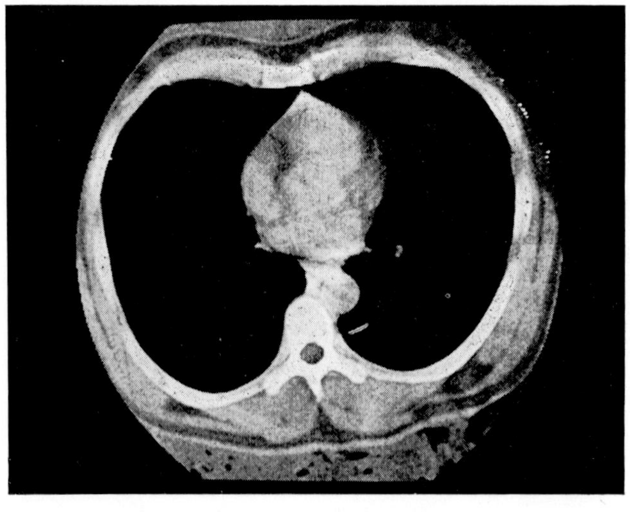

(*b*)

Fig. 6.40 (*cont. opposite*).

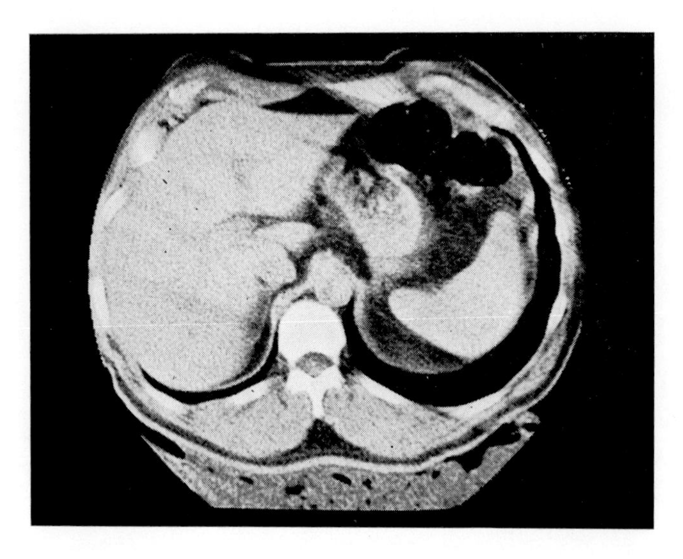

(*c*)

Fig. 6.40. Schematic diagram of the technique used to produce X-ray sections. (*b*) Section through chest showing bone, muscle and fat with the heart as the central feature. (*c*) Section through abdomen showing the liver on the left, the pancreas in the middle and the spleen on the right. By permission of E.M.I. Central Laboratories.

Fig. 6.41 (*a*) shows a pattern of four spots with six directions indicated. Fig. 6.41 (*b*) shows ' radiographs ' using two of these directions. Fig. 6.41 (*c*) shows a ' smear ' pattern derived from (*b*); the grey lines represent loci of possible positions that could be deduced from the radiographs. The intersections are locations that would fit both radiographs. Fig. 6.41 (*d*) shows the process applied to six directions and the intersections now give unambiguous locations for all four spots. Clearly a larger number of directions of projection would greatly increase the resolution. The computer, in effect, superposes the data from the large number of directional projections in precisely the same way.

At first this seems to be a totally new method of imaging, but in fact it does conform to our principles. We included scanning as a direct recombination process using point-by-point assembly because usually —as for television for example—the process is so rapid that the appearance to an observer is the same as if the whole picture had been imaged simultaneously as with a lens. In this example we are scanning —but it would be more correct to describe this as an indirect recombination process. The scanning here is ' angle-by-angle '. Each X-ray

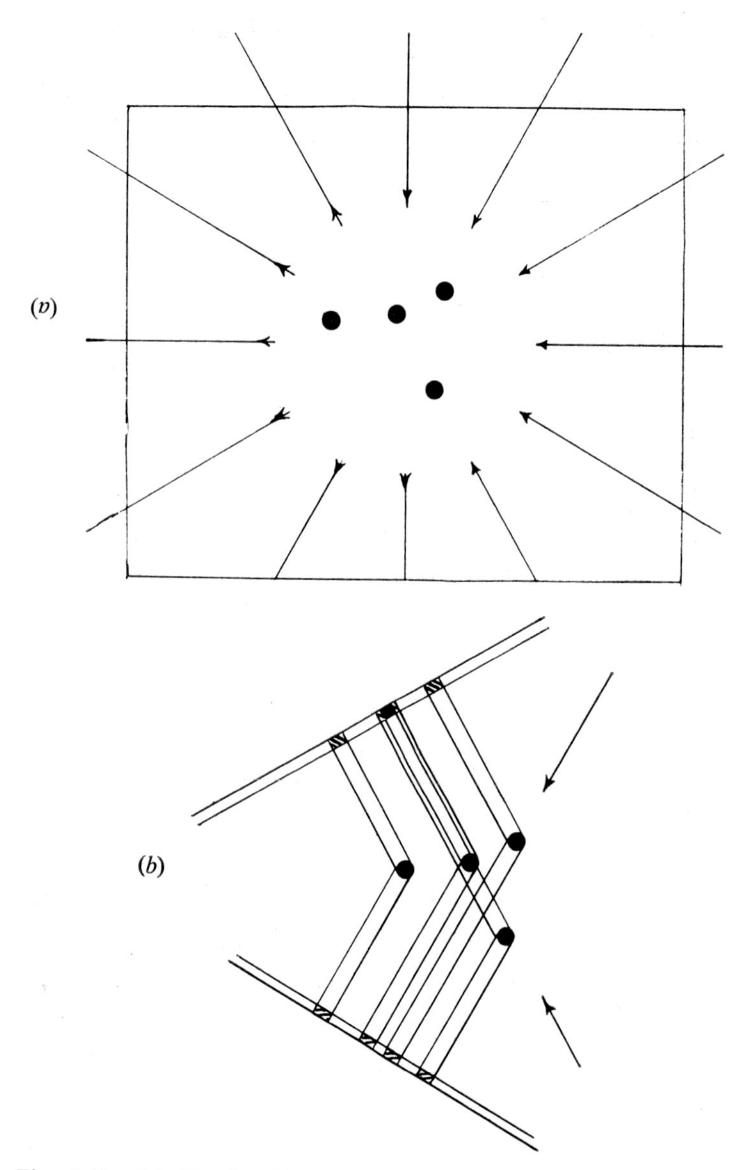

Fig. 6.41. (*a*) Simple object of four points with six viewing directions indicated. (*b*) Projections of the object along two of the chosen directions. *Opposite:* (*c*) Superposition of two patterns which represent possible locations of the points of (*a*) derived from the projections (*b*). (*d*) Superposition of 6 patterns, each derived from a projection in one of the directions indicated in (*a*): the positions of the original points are clearly revealed. With very much greater numbers of projections the precision would obviously increase.

(c)

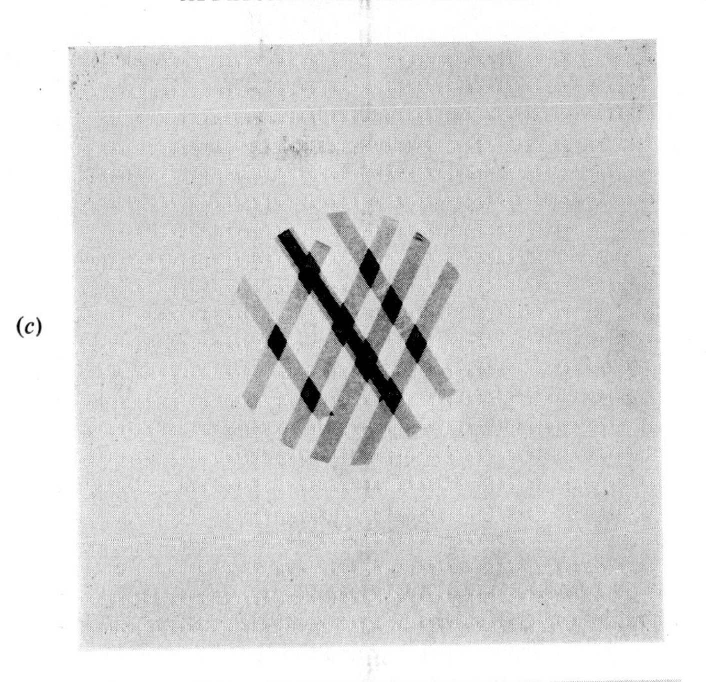

(d)

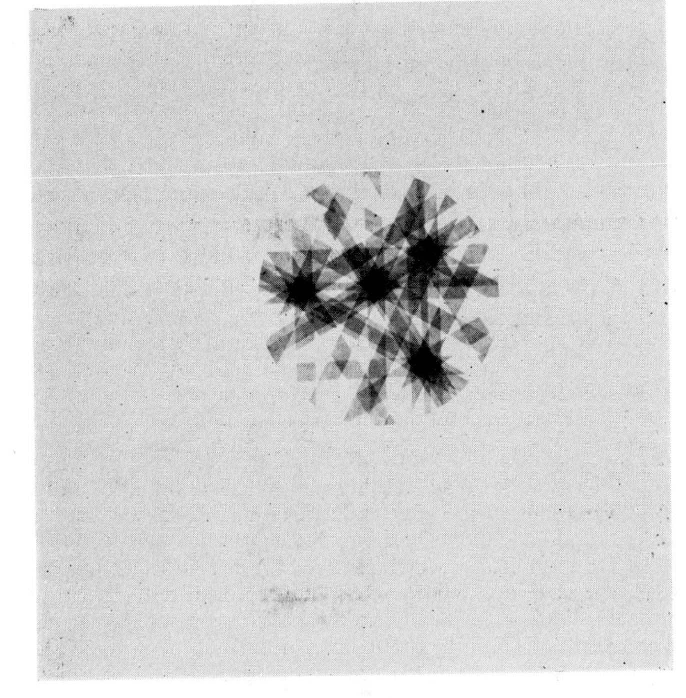

photograph is taken at a different angle and each can be regarded as giving a block of information. The big difference is that we cannot present the result without first combining this information by means of a computer program. The 'known fact' that enables us to do the recombination or focusing is that every X-ray photograph in all the different directions corresponds to a single object which is the same for each.

The dangers associated with exposure of human tissue to X-radiation have become increasingly worrying and various techniques have been tried out as a substitute. Perhaps the most successful is that of ultrasonic scanning, which has proved particularly valuable in examinations during pregnancy. The technique is simple in principle—though realization in practice presents problems, largely because of the difficulty in achieving sufficiently high levels of radiation. In early systems the patient lay on a table over a tank of water with the portion of the body to be scanned in contact with the water surface through a hole in the table (fig. 6.42 (*a*)). An ultrasonic generator directs a beam of ultrasonic radiation through the water on to the patient and the scattered radiation is picked up by a suitable receiver in the water. The incident beam scans the patient and the resulting picture is built up point by point. The system is rather like a miniature radar system; for any given direction of incidence there will be reflections from all the interfaces between different materials, e.g. between water and skin, between flesh and bone, between the wall of the uterus and the fluid inside, etc. The reflections arise because the velocity of ultrasonic waves is different in the different media and so in effect the refractive index changes. Thus for each direction of the beam, a series of spots on the presentation screen will be produced and, as the angle of incidence is changed to scan the patient, the picture will be built up. Modern systems are simpler; the transmitter/receiver is held in contact with the body using a liquid lubricant to ensure acoustic contact. Figs. 6.42 (*b*) and (*c*) show examples of results. Fig. 6.42 (*c*) is a

Opposite:

Fig. 6.42. (*a*) Diagram of an early ultrasonic scanning system. (*b*) Ultrasonic scan of the modern 'gray scale' type in which variations of absorption show up as variations in the gray tone: this scan is of a patient lying on her back, scanned longitudinally along the centre line and shows clearly the skull of the baby in her womb (marked fs) and its trunk (marked ft). (*c*) 'Echo' type scan in which the white lines represent boundaries between various tissues, liquids, cavities, etc.: the patient is again horizontal and scanned at a slight angle to the centre line. The skulls of twins can be seen very clearly. By permisison of Picker Corporation.

(*a*)

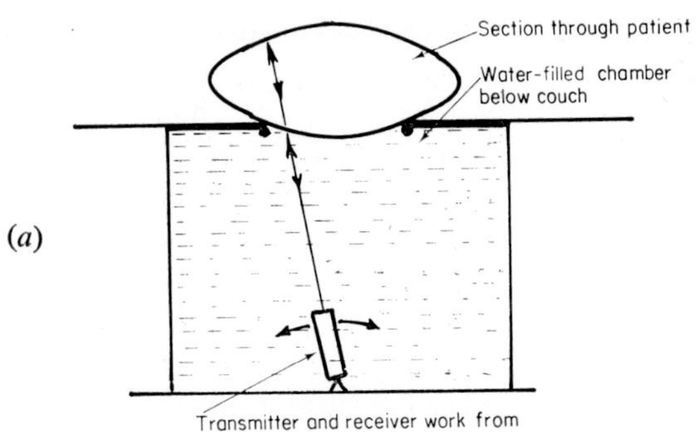
Section through patient
Water-filled chamber below couch
Transmitter and receiver work from side to side and travel perpendicular to the drawing in order to complete scan

(*b*)

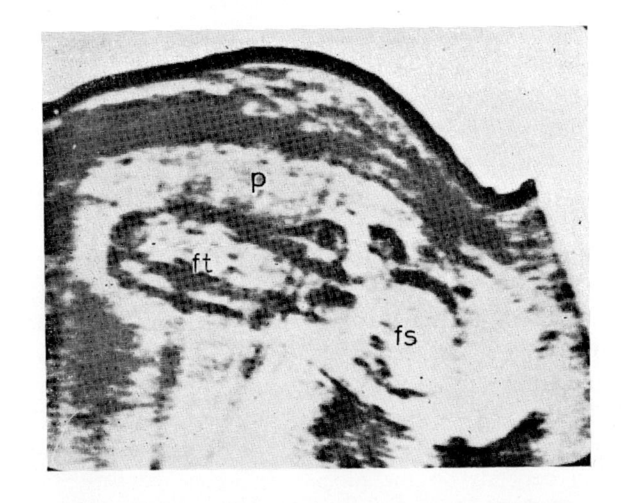
p
ft
fs

(*c*)

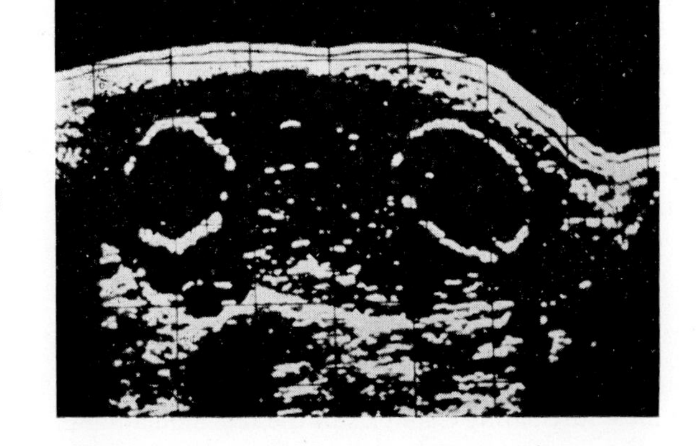

particularly striking example in which the reflection from the heads of twins inside the womb can clearly be seen. This is a particularly good example of our process of scattering and recombination in which the recombination is by direct point-by-point assembly.

For both X-rays and ultrasonics we irradiated the object and collected the scattered radiation. We shall now turn to two examples in which the objects are ' self-luminous '.

The first is of growing importance in diagnosing many different diseases and corresponds to the use of radioactive tracers in industrial applications. The patient either ingests or is injected with a chemical which includes a radioactive isotope. The chemical is chosen so that it will be concentrated by the body in the organ to be examined (e.g. radio-iodine concentrates in the thyroid gland) and of course the half-life of the isotope and the dose are selected to ensure minimal danger from the radiation. The result however is that the body contains a ' self-luminous ' organ emitting gamma radiation. In order to recombine the radiated beams to produce an image, a scanning process is used which involves the principle of ' straight-line ' imaging. In principle a long, fine hole in a massive tube, say of lead, will allow the passage of gamma rays in only one direction. By directing the tube to different points of the body in turn a picture of the organ concerned can be built up. It is usual nowadays to automate the system so that the radiation received by the tube in any direction is recorded by a scintillation counter and the whole record is processed by a small computer and presented, usually in colour, on a television type screen.

Finally in this section, it is worth mentioning thermography. The human body radiates infrared radiation the whole time, but the amount of radiation depends on the precise difference between the surface temperature and the surroundings. In various diseases, the temperature of the body can be slightly changed and hence an image of the body built up from its ' self-luminous ' infrared radiation can be useful in diagnosis.

Fig. 6.43 (*a*) shows a schematic diagram of a thermographic system. It again involves recombination of the radiated beams by the point-by-point scanning process. Fig. 6.43 (*b*) shows a typical example of a thermogram. More recent systems use what is, in effect, an infrared television camera, and by electronic treatment of the signals, can display temperature contours as colour variations.

6.13. *Holography*

The principles of holography were outlined in Section 4.3 and provide one of the most fascinating examples of the recombination of scattered

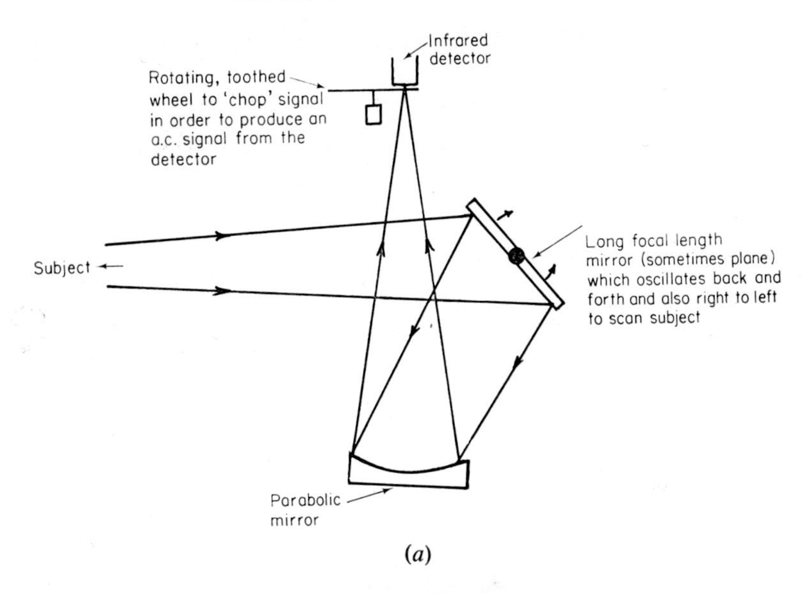

(*a*)

(*b*)

Fig. 6.43. (*a*) Schematic diagram of thermograph system. (*b*) Thermogram of patient with an advanced cancer of the left breast. The dark areas indicate the high temperature regions. This photograph is really only illustrative of the principle as such an advanced cancer would normally have been detected by other means. By permission of Dr. D. K. L. Davies and Mr. R. E. Toogood, Velindre Hospital, Cardiff.

radiation to produce an image. Our intention here is merely to describe a few practical examples of applications of holography.

The first, which I include because it is of particular interest to me, is to the testing of components of musical instruments of the string family. Violins, guitars, etc. all involve two major components, the strings which determine the frequency of the note being produced and its relative amplitude and duration, and the body which amplifies the sound and in so doing modifies its quality very considerably. The vibrational response of the back and belly plates are thus crucially important in determining the tone quality. The problem is, however, that it is not too easy to determine the resonant response during construction while it is still possible to make modifications; nor is it easy to know what modifications to shape, thickness, etc. are needed in order to make a desired change in vibrational characteristics. One solution that has been used is to excite the plate with an electromechanical driving unit and to scatter sand on it: the sand collects along nodal lines to form the well known Chladni figures and these in turn enable deductions about the characteristics to be made. The amplitude of excitation required is however very high—much higher than that used when the instrument is played. Hence the results may not be meaningful in relation to practical playing problems. Holography provides an answer.

We set up the violin plate and produce a hologram of it. The system is then set up again with the *real* violin plate coinciding precisely with the *image* of the violin plate recombined by irradiating the hologram with a laser beam. If the two do not coincide *precisely* interference figures are seen and if the real violin plate is driven by a quite weak oscillator the interference fringes show up the vibrational pattern. Fig. 6.44 shows examples of this technique in use.

The second example is a particularly elegant technique which is a very appropriate one with which to end a book on images, since it uses one of the most modern image-forming techniques to produce special lenses which themselves are to be used for image forming. With highly coherent laser light, conventional lenses scatter so much that internal diffraction problems produce insurmountable difficulties and holographic lenses may be the solution to this problem.

The principle is based on that of the zone plate (Section 3.4) and the hologram (Section 4.3). The problem is first to provide a zone plate with sinusoidally varying zones so that it will only have one focal length—a point which was explained in Section 4.3. Then we need to improve the transmission—even a sinusoidally varying zone plate still works by discarding half the incident light. The solution is to

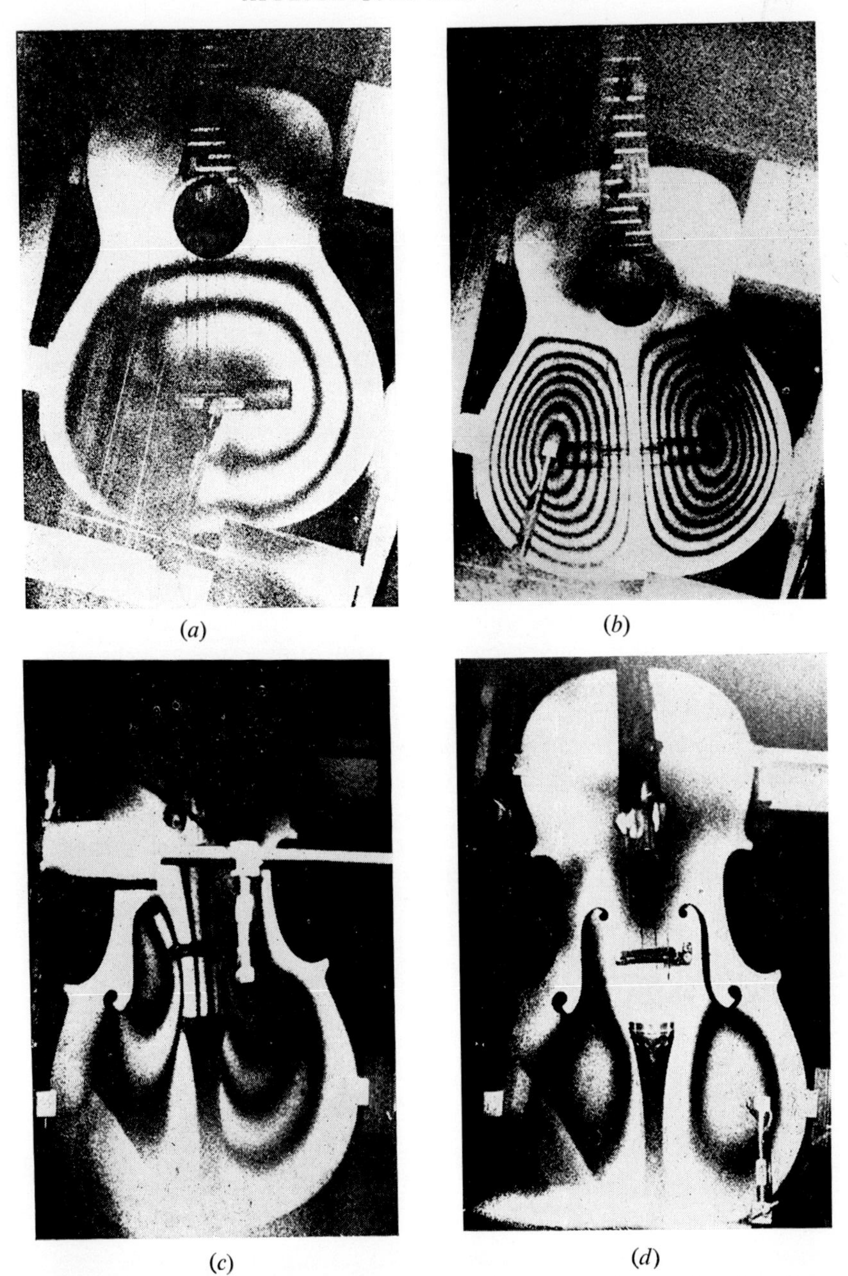

(*a*) (*b*)

(*c*) (*d*)

Fig. 6.44. Reconstructed holograms showing two modes of vibration for guitar and cello front plates. The frequencies of excitation are (*a*) 148 Hz, (*b*) 236 Hz, (*c*) 174 Hz and (*d*) 292 Hz. By permission of Dr. Ian Firth, Department of Physics, University of St. Andrews.

use a *phase* zone plate in which the light waves transmitted through the various annuli have their phases changed so that all may contribute constructively. How then does this differ from an ordinary lens? Is not this what a conventional lens does? The answer is " Not quite ". The conventional lens adds various additional path lengths in order to bring all paths through the lens to be exactly equal when an image is in focus. The phase zone plate keeps the paths unequal but ensures that they always differ by a whole number of wavelengths; optically speaking therefore no problem arises. The great advantage is that the phase zone plate can be extremely thin even if the aperture is very large, and this reduces background scattering, especially when used in coherent light.

But how can such a plate be made? The practical details are complex but the principle is simple. In our discussion on holography (Section 4.3) the idea of the interaction between a spherical wave diverging from a point and a plane reference beam giving rise to a zone-plate pattern was used. If that pattern is recorded not on a photographic emulsion which gives a black-and-white reproduction but in a photo-sensitive polymer which becomes soluble to a greater or lesser extent in a suitable solvent depending on the amount of exposure to light, it is possible to produce a transparent sheet of polymer with exactly the desired variations in thickness. This method of producing holographic lenses has only arrived at a usable level of perfection very recently but is likely to revolutionize laser optics in the near future.

There is considerable aesthetic satisfaction in seeing an argument come round into a full circle and in being able to use in a very practical and useful way a principle which, when it was originally introduced, was regarded as little more than a fascinating scientific curiosity.

Further reading

The general fields of optical and X-ray diffraction and of image formation are covered by a bewildering number of books at many different levels. This brief selection is intended to give a few starting points for those wishing to delve more deeply into the subjects. They are arranged in chronological order.

TAYLOR, C. A., and LIPSON, H. (1964). *Optical Transforms.* London: G. Bell & Sons.

PAMENT, G. B., and THOMPSON, B. J. (1969). *Physical Optics Notebook.* California: Society of Photo-optical Instrumentation Engineers.

LIPSON, G., and LIPSON, H. (1969). *Optical Physics.* Cambridge: The University Press.

LIPSON, H. (1970). *Crystals and X-rays.* London: Wykeham Publications.

WOOLFSON, M. M. (1970). *X-ray Crystallography.* Cambridge: The University Press.

SMITH, F. G., and THOMSON, J. H. (1971). *Optics.* London: Wiley.

WILLIAMS, C. S., and BECKLAND, O. A. (1972). *Optics, A Short Course for Engineers and Scientists.* New York: Wiley.

LIPSON, H. (Editor) (1972). *Optical Transforms.* London: Academic Press.

HARBURN, G., TAYLOR, C. A., and WELBERRY, T. R. (1975). *Atlas of Optical Transforms.* London: G. Bell & Sons.

Images and Information. Open University Course Text, 1977.

Index

THE WYKEHAM SCIENCE SERIES

1	*Elementary Science of Metals*	J. W. MARTIN and R. A. HULL
2	†*Neutron Physics*	G. E. BACON and G. R. NOAKES
3	†*Essentials of Meteorology*	D. H. MCINTOSH, A. S. THOM and V. T. SAUNDERS
4	*Nuclear Fusion*	H. R. HULME and A. McB. COLLIEU
5	*Water Waves*	N. F. BARBER and G. GHEY
6	*Gravity and the Earth*	A. H. COOK and V. T. SAUNDERS
7	*Relativity and High Energy Physics*	W. G. V. ROSSER and R. K. MCCULLOCH
8	*The Method of Science*	R. HARRÉ and D. G. F. EASTWOOD
9	†*Introduction to Polymer Science*	L. R. G. TRELOAR and W. F. ARCHENHOLD
10	†*The Stars; their structure and evolution*	R. J. TAYLER and A. S. EVEREST
11	*Superconductivity*	A. W. B. TAYLOR and G. R. NOAKES
12	*Neutrinos*	G. M. LEWIS and G. A. WHEATLEY
13	*Crystals and X-rays*	H. S. LIPSON and R. M. LEE
14	†*Biological Effects of Radiation*	J. E. COGGLE and G. R. NOAKES
15	*Units and Standards for Electromagnetism*	P. VIGOUREUX and R. A. R. TRICKER
16	*The Inert Gases: Model Systems for Science*	B. L. SMITH and J. P. WEBB
17	*Thin Films*	K. D. LEAVER, B. N. CHAPMAN and H. T. RICHARDS
18	*Elementary Experiments with Lasers*	G. WRIGHT and G. FOXCROFT
19	†*Production, Pollution, Protection*	W. B. YAPP and M. I. SMITH
20	*Solid State Electronic Devices*	D. V. MORGAN, M. J. HOWES and J. SUTCLIFFE
21	*Strong Materials*	J. W. MARTIN and R. A. HULL
22	†*Elementary Quantum Mechanics*	SIR NEVILL MOTT and M. BERRY
23	*The Origin of the Chemical Elements*	R. J. TAYLER and A. S. EVEREST
24	*The Physical Properties of Glass*	D. G. HOLLOWAY and D. A. TAWNEY
25	*Amphibians*	J. F. D. FRAZER and O. H. FRAZER
26	*The Senses of Animals*	E. T. BURTT and A. PRINGLE
27	†*Temperature Regulation*	S. A. RICHARDS and P. S. FIELDEN
28	†*Chemical Engineering in Practice*	G. NONHEBEL and M. BERRY
29	†*An Introduction to Electrochemical Science*	J. O'M. BOCKRIS, N. BONCIOCAT, F. GUTMANN and M. BERRY
30	*Vertebrate Hard Tissues*	L. B. HALSTEAD and R. HILL
31	†*The Astronomical Telescope*	B. V. BARLOW and A. S. EVEREST
32	*Computers in Biology*	J. A. NELDER and R. D. KIME
33	*Electron Microscopy and Analysis*	P. J. GOODHEW and L. E. CARTWRIGHT
34	*Introduction to Modern Microscopy*	H. N. SOUTHWORTH and R. A. HULL
35	*Real Solids and Radiation*	A. E. HUGHES, D. POOLEY and B. WOOLNOUGH
36	*The Aerospace Environment*	T. BEER and M. D. KUCHERAWY
37	*The Liquid Phase*	D. H. TREVENA and R. J. COOKE
38	†*From Single Cells to Plants*	E. THOMAS, M. R. DAVEY and J. I. WILLIAMS
39	*The Control of Technology*	D. ELLIOTT and R. ELLIOTT
40	*Cosmic Rays*	J. G. WILSON and G. E. PERRY
41	*Global Geology*	M. A. KHAN and B. MATTHEWS
42	†*Running, Walking and Jumping: The science of locomotion*	A. I. DAGG and A. JAMES
43	†*Geology of the Moon*	J. E. GUEST, R. GREELEY and E. HAY
44	†*The Mass Spectrometer*	J. R. MAJER and M. P. BERRY
45	†*The Structure of Planets*	G. H. A. COLE and W. G. WATTON
46	†*Images*	C. A. TAYLOR and G. E. FOXCROFT
47	†*The Covalent Bond*	H. S. PICKERING
48	†*Science with Pocket Calculators*	D. GREEN and J. LEWIS

THE WYKEHAM ENGINEERING AND TECHNOLOGY SERIES

1	*Frequency Conversion*	J. THOMSON, W. E. TURK and M. J. BEESLEY
2	*Electrical Measuring Instruments*	E. HANDSCOMBE
3	*Industrial Radiology Techniques*	R. HALMSHAW
4	*Understanding and Measuring Vibrations*	R. H. WALLACE
5	*Introduction to Tribology*	J. HALLING and W. E. W. SMITH

All orders and requests for inspection copies should be sent to the appropriate agents. A list of agents and their territories is given on the verso of the title page of this book.

†(*Paper and Cloth Editions available.*)